General Pharmacology

WHO Definition of Drug *"Any substance or product that is used or intended to be used to modify or explore physiological systems or pathological states for the benefit of the recipient"*

Sources of Drugs

1. **Natural Sources**
i. **Vegetable Sources:** They are obtained from Plants
   a) **Alkaloids:** Nitrogenous heterocyclic bases derived from plants.
   ✓ They are insoluble in water but forms salts with water which are water soluble
   ✓ Eg: Atropine from *Atropa belladonna*, Quinine from *Cinchona bark*, Morphine from *Papavarum somniferum*, Reserpine from *Rauwolfia serpentine*, Nicotine from *Tobacco leaves*
   b) **Glycosides:** in these compounds a sugar moiety is joined to a non-sugar moiety with a ether linkage (-O-)
   ✓ The pharmacological action lies in its non-sugar residues (aglycon)
   ✓ The sugar portion guides the pharmacokinetic characteristics
   ✓ Eg: Cardiac Glycosides used in CHF (Digitoxin, Digoxin obtained from Fox Glove leaves (*Digitalis Purpurea*))
   c) **Oils:**

Essential Oils (or Volatile oils)	Fixed Oils	Mineral Oils
• They are obtained from leaves or flowers • They are used as Carminatives, Astringents in Mouth washes or as Flavoring agents eg: Eucalypus Oil, Clove Oil, Peppermint Oil, Ginger Oil	• These are non volatile oils. • They are obtained from Seeds • Eg: Groundnut Oil, Coconut Oil, Olive Oil	• These are petroleum products and obtained from dry distillation of wood. • These are mainly used as vehicles for preparation of ointments • Eg: Hard and Soft paraffin

   d) **Tannins:** These are non-nitrogenous phenolic derivatives from plant sources and are soluble in water
   ✓ They are mainly used a Astringents
   ✓ Eg: Tinct. Catechu, Tinct. Rhubarb
   e) **Gums:** These are colloidal exudates of he plants
   ✓ These are mainly used as emulsifying or suspending agents
   f) **Resins:** These are polymers of volatile oil and are insoluble in water
   ✓ Eg: Benzoin (used as inhalation in common cold), Tinct.Benzoin (as antiseptic protective sealing over bruises)
ii. **Animal Sources:** They are obtained from living or dead animals or from their parts
   ✓ These comprise hormones, Vitamins, Vaccines and Sera
   ✓ Eg: Insulin from Pancreas, Vit B_{12} from Liver Extract, Thyroxin from Thyroid
iii. **Microbial Sources:** Mostly they are obtained from Fungi, Moulds & Bacteria
   ✓ These are mostly life saving drugs
   ✓ Eg: Penicillin from *Penicilium notatum*, Chloramphenicol from *Streptomyes venezuelae*, Griseofulvin from *Penicilium griseofulvum*
iv. **Mineral Sources:** They are naturally available minerals used for therapeutic purpose
   ✓ Eg: Ferrous Sulfate (in anemia), Magnesium Sulfate (as Purgative), Aluminium Hydroxide & Sodium Bicarbonate (as antacids), Radioactive isotopes like I^{131} for the diagnosis & treatment of Thyrotoxicosis & Thyroid malignancy and P^{32} for Polycythaemia vera
2. **Semi-Synthetic Sources:** The naturally obtained dreg has been chemically modified in laboratory for better pharmacokinetic profile and also to reduce the toxicity.
Example: Desferrioxamine – Antidote for Iron poisoning. Ferrioxamine is isolated from *Streptomyces pilosus* and degraded chemically to obtain iron free ligand Desferrioxamine.

3. **Synthetic Sources:** These drugs are completely synthesized in the laboratory

✓ Majority of the drugs are synthetic drugs

✓ They are having better efficacy, efficiency, and less toxic effects compare with naturally obtained drugs

✓ Their chemical structure can be modified for getting better speicificity

✓ Eg; Aspirin, Paracetamol

4. **Genetically Engineered Drugs:** These are obtained from new methodology involves the blending of discoveries from Molecular biology, Recombinant DNA technology, DNA alteration, Gene Splicing, Immnology & Immuno Pharmacology

✓ Eg: Recombinex HB – a hepatitis-B vaccine, Humulin – human insulin of recombinant DNA origin, Interferon-alpha-2a & Interferon-alpha-2b for hairy cell leukaemia

Routes of Drug Administration

I. **Enteral Routes:** Placement of drug directly into any part of the GIT is called an 'enteral' mode of administration

a) **Oral:** Swallowing a drug through mouth

Advantages	Disadvantages
✓ Most commonly used method as it is safe, convenient & painless procedure ✓ Economical as sterilization of drug products is not essential ✓ No need of any assistant	✓ Onset of action is slower ✓ Polar drugs can't be given as they are not absorbed (eg: Streptomycin) ✓ Drugs are destroyed by the digestive juices (Eg: Penicillin-G, Insulin, Oxytocin) ✓ 1st pass effect (those destroyed in liver before reaching systemic circulation) (eg: Morphine, Isoprenaline) ✓ Bad test & Bad smell & irritant drugs can't be given ✓ Drugs can't be given to unconscious & uncooperative patients ✓ Drugs can't be given during emesis

b) **Sublingual / Buccal**

✓ The drug is place beneath the tongue (sublingual) or crushed in mouth and spread over the buccal mucosa (Buccal).

Advantages	Disadvantages
✓ Quick onset of action because of rapid absorption due to more blood supply in that region ✓ Bypasses the portal circulation → no 1st pass metabolism ✓ Drug action can be terminated at any time when side effects are observed	✓ Distasteful, irritant drugs can't be given ✓ Higher molecular weight drugs can't be absorbed (eg; insulin)
Examples	
Isosorbide dinitrate tablets & Nitroglycerin tablets (for Angina), Isopranline sulfate tablets (for Bronchial Asthma), Nifedipine in powder form (in Hypertension)	

c) **Rectal:** Through Rectum (Suppositories, Enema)

Advantages	Disadvantages
✓ Useful in patients with nausea and vomiting ✓ 1st pass metabolism is greatly bypassed as a major portion of the drg is absorbed from external haemorrhoidal veins ✓ Useful for gastric irritant drugs	✓ Chances of rectal inflammation ✓ Absorption is irregular ✓ Inconvenient and embarrassing to the patient

> *Examples*
>
> ✓ Dulcolax & Glycerine suppositories, enemas, ointments for Local action
> ✓ Aminophylline (Bronchodillator) & Indomethacin (Anti-inflammatory agent) Suppositories for Systemic action

II. **Parenteral Routes:** Routes other than "Enteral" are called 'Parenteral' routes of administration. Administration of drugs by injection, by topical application to skin or by inhalation through the lungs are all parenteral.

    a) **Intravenous:** Through lumen of the veins

Advantages	Disadvantages
✓ Directly enters into the systemic circulation → no 1st pass effect & quicker onset of action ✓ Less dose is needed to achieve greater therapeutic effects ✓ Valuable in emergency ✓ Can be given evening unconscious, uncooperative patients those are having nausea, vomiting & diarrhea ✓ Hypertonic solutions & GIT irritant drugs can be infused ✓ Large volume of fluids can be infused at a uniform rate ✓ Amount of the drug can be controlled with an accuracy	✓ Strict aseptic conditions are needed ✓ Patient has to depend upon other person for administration of drug ✓ Painful ✓ Risky because once the is injected it can't be recalled ✓ Introduction of any air or particulate matter produce embolism which is fatal ✓ Drugs in suspensions & Oily drugs can't be given ✓ Depot injections can's be given ✓ Venous thrombosis & Thrombophlebitis of the vein injected ✓ Necrosis around the site of action
Examples: Glucose, Glucose normal saline, Dopamine & Norepinephrine drips	

    b) **Intramuscular**
    ✓ Deltoid muscle or gluteal mass of left or right buttock
    ✓ Vastus muscle underlying the lateral surface of the thigh

Advantages	Disadvantages
✓ Absorption is more predictable, less variable & rapid compared to Oral route ✓ Depot injections can be given	✓ Perfect aseptic conditions are needed ✓ Chances of abscess at the site of injection ✓ Chances of nerve damage leading to paresis of muscle supplied by it ✓ Large volumes can't be given (maximum 5 – 10 ml)
Examples: Depot injection of Testosterone, Antibiotics, Antiemetics	

c) **Intraperitoneal**	d) **Intrathecal (Intraspinal)**
✓ Into the peritoneal space ✓ Rapid absorption due to large surface area ✓ Painful, risky ✓ Antirabies injection can be given	✓ Into the subarachnoid space ✓ They crosses BBB & Blood CSF barrier ✓ Strict aseptic conditions & grater expertise is needed ✓ Its painful & risky procedure ✓ Many radiopaque contrast media for myelography (to visualize spinal cord) are given through this route ✓ Xylocaine injection for providing Spinal Anesthesia
e) **Intramedullary:** Injection into the tibial or sternal bone marrow	f) **Intra-arterial:** into the lumen of the desired artery

g) **Intra-articular:** injection directly into the joint space	h) **Subcutaneous:** Injection into the subcutaneous tissue under the skin
i) **Inhalation:** Inspiration through nose or mouth	

III. **Topical Routes**
   a) **Transdermal**
   ✓ Transdermal Patches
   ✓ In these adhesive patches, the drug is incorporated into a polymer (usually Polyisobutylene) which in turn is bonded to an adhesive plaster
   ✓ The drug is delivered at the skin surface by diffusion, for percutaneous absorption into circulation
   ✓ These preparations are designed to provide steady & smooth plasma concentration of the drug for a period ranging from 1-3 days from the site of their application
   ✓ Site of application: Chest, Abdomen, Upper arm or Mastoid region
   ✓ Examples: Transdermal Patches of Nitroglycerine, Scopalamine, Clonidine, Estradiol
   ✓ *For diagram refer KD Tripathi Text book*
   b) **Conjuctival:** into the conjuctiva for local effects eg: Sulfacetamide
   c) **Vaginal and Urethral:** Pessaries are used for local actions
   d) **Inunction (Rubbing):** rubbing onto the skin

Newer Drug Delivery Systems: To improve drug delivery and to prolong its duration of action, special drug delivery systems have recently been developed.

   ✓ *These include:* Ocuserts, Progestaserts, Transdermal Adhesive Patches, Prodrugs, Computerised Miniature Pumps, Use of Monoclonal Antibodies and Liposomes as drug carriers.
Dosage Forms

Solid Dosage Forms: Tablets, Capsules, Powders, Effervescent Powders, Granules

- **Tablets:** These are powdered or granulated form of drug compressed under heavy pressure into a round or disk like shape suitable for swallowing
- **Sugar coated tablets:** tablets coated with sugar film to avoid bitter taste
- **Film coated tablets:** coated with film containing gelatin or cellulose to mask the taste
- **Enteric coasted tablets:** coated with cellulose acid phthalate, shellac → coating is resistant to gastric acid but dissolves at intestinal alkaline pH → prevents the degradation at gastric pH
- **Sustained release tablets:** Aggregated drug particles have individual coating with different type of inert resins → each type of coating dissolves at different time intervals → action will be there for 10-12 hrs
- **Pellets:** Sterile spheres by compression of drug powder which are implanted subcutaneously & form a depot → drug is slowly released for a long duration of time
- **Lozenges:** tablet containing drug with sugar & a gum and is meant for chewing or sucking for providing local effects in mouth or throat
- **Capsules:** tasteless gelatin containers, of appropriate size, having powdered drug & excipients → meant for swallowing
- **Spanules:** longer acting capsules
- **Powders:** simple or mixture of drugs in a dried & finely pulverized form intended for external use or for oral use
- **Effervescent powders:** Powdered drugs mixed with sodium bicarbonate, citric acid or tartaric acid → when dissolved in water → they effervesce with evolution of CO_2 → to make the mixture more palatable

- **Granules:** Small aggregates of powder held together by a binding agent → can be dissolved in a specified volume of water to make a suspension for immediate oral use in children

Liquid Dosage Forms: Aqueous Solutions(Syrups, Liquors, Linctus, Injections), Aqueous Suspensions (Mixtures, Emulsions), Alcoholic Solutions (Spirits, Elixirs), Drops, Extractives, Enema

- **Syrups:** Drugs in concentrated solution of sugar plus favouring agents & permitted colors
- **Liquors:** Aqueous solutions of medical substances which are either gases or volatile or which sublimate
- **Linctus:** viscous syrupy liquids containing the drug with some demulcent like menthol → these are to be sipped without dilution to provide soothing effect in sore throat
- **Injections:** Sterile solutions or sterile suspensions of the drug in suitable solvent plus preservatives & are meant for parenteral use
- **Depot Injections:** longer acting injectable preparation
- **Mixtures:** solid drugs homogeneously in water by means of some harmless suspending agent
- **Emulsions:** two or more immiscible liquid medicaments are dispersed by means of some suspending agent
- **Spirit:** 10% v/v solution of volatile essential oils plus alcohol
- **Elixirs:** pleasantly flavoured solutions of a drug in sugar syrup or glycerol along with higher proportions of alcohol
- **Drops:** mainly paediatric preparations which contain small amounts of highly concentrated solutions of drugs
- **Extractives:** made from vegetable drugs & contain the active principles of crude drug in an alcoholic or hydro-alcoholic solution
- **Enema:** medicated liquid preparations for rectal administration & are used for evacuation of colon

Dosage forms for external use

- **Liniments:** liquid medicaments to be rubbed on skin with friction
- **Lotions:** liquid medicaments for local applications but without rubbing
- **Ointments:** soft, semi-solid masses containing the drug in a greasy base, like soft or hard paraffin
- **Paste:** like an ointment but it doesn't have a gresy base
- **Inhalants:** Liquid preparations containing a drug to be inhaled as vapour
- **Aerosols:** the drug dissolved in a liquid is put inside a cylindrical container & then filled with a propellant gas under pressure → a push at the valve releases the drug through a microfined orifice in the form of mist which is inhaled
- **Suppositories:** For rectal use
- **Pessaries:** for vaginal use
- **Bougies:** for urethral use
- Transdermal Adhesive Patch

Pharmacokinetics

The study of time course of drug ADME & their relationship with its therapeutic & toxic effects of drug (What Body does to the drug) – Quantitative aspects of drug

ABSORPTION: The process of movement of drug from its site of administration to the systemic circulation

DISTRIBUTION: The movement of drug between one compartment & the other

ELIMINATION: The process that tends to remove the drug from the body & terminates its action.

Includes both Biotransformation & Excretion

Absorption

➤ The process of unchanged drug from the site of administration to systemic circulation.
➤ Effectiveness of a drug → assessed by its conc. At the site of action
Minimum Effective Conc. : The minimum conc required to show the action by the drug

Minimum Toxic Conc. : The minimum conc required to show the toxic effects by the drug → Maximum Effective Conc.

MECHANISMS OF DRUG ABSORPTION

1)   PASSIVE DIFFUSION   2) PORE TRANSPORT OR FILTRATION   3)   FACILITATED DIFFUSION 4) ACTIVE TRANSPORT   5) ENDOCYTOSIS
Passive Diffusion

- Non-Ionic diffusion
- Major process for absorption of more than 90% of drugs.
- Driving force → concentration or electrochemical gradient
- Drug movement is result of KE of molecules.
- No energy source is required → PASSIVE
- Down Hill transport
- It is expressed by FICK's First law of diffusion *"The drug molecules diffuse from a region of higher concentration to one lower concentration until equilibrium is attained & the rate of diffusion is directly proportional to the conc gradient across the membrane."*
- Greater area & Lesser thickness of membrane → Faster the diffusion.
- Rate of transfer unionized species is 3-4 times the rate for ionized drugs
- Greater the membrane / water partition coefficient of drug → Faster the absorption
- Drug diffuses rapidly → when volume of GI fluid is low.
Pore Transport or Filtration

- Imp in absorption of low molecular weight (<100D), low molecular size e.g: Urea, Sugars
- Driving Force → *hydrostatic pressure or the osmotic differences across the membrane due to which bulk flow of water along with small solid molecule occurs through such aqueous channels*

- This is imp in renal excretion, removal of drug from CSF & entry of drugs into the liver

Carrier Mediated Transport

- Involve a component of membrane called "Carrier" that binds reversibly or

non-covalently with solute molecules to be transported.

- This carrier returns back; Carrier may be enzyme or some other component of membrane

- Carriers have structure specificity → competition
- At higher drug conc the system becomes saturated
- An area in which the carrier system in most dense
    - → Absorption Window

- 2 types → Facilitated Diffusion, Active Transport

Active Transport

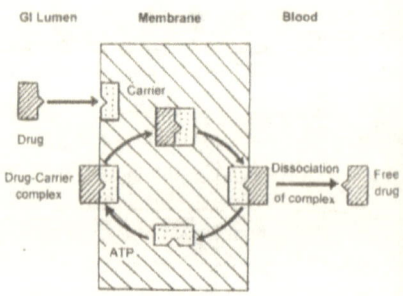

- Drug is transported from a region of lower to one of higher conc i.e. against conc gradient or uphill transport

without any regard for equilibrium.

- Up Hill process → Energy required in work done by carrier
- It can be inhibited by metabolic poisons that interfere with energy production like Fluorides, Cyanide & Dinitrophenol & lack of O2

Examples

- Endogenous substances like Sodium, Potassium, Calcium, Iron, Glucose, Certain Amino Acids, Vitamins like Niacin, Pyridoxine, Ascorbic Acid

- Absorption of 5-FU via Pyrimidine Transport System
- Absorption of ACE inhibitor Enlapril via small peptide carrier system
- Active transport imp in Renal & Biliary excretion of many drugs & their metabolites & secretion of certain acids out of the CNS.

IONIC OR ELECTROCHEMICAL

- Charge on the membrane influences the permeation of drugs

- Anionic solute permeates faster than the Cationic form

- Permeation of Cation depend upon potential

difference or electrical gradient as the driving force across the membrane

- Cations repelled due to +ve charge on outside the membrane → so cations with ↑ K.E → penetrate ionic barrier → inside the membrane cation + Anion → electric gradient → downhill → Electrochemical Diffusion

<u>Endocytosis</u>

- Minor transport mechanism which involves engulfing extra cellular materials within a segment of cell membrane to form

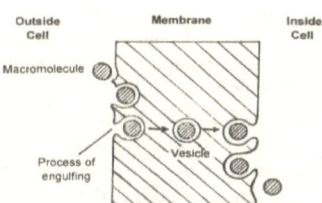

a saccule or a vesicle which is then pinched off intracellularly.

- Drug is absorbed into lymphatic circulation → by passing First Pass Metabolism

- Eg: Fats, Starch, Oil, A, D, E, K, Insulin
- Polio Vaccine → Pinocytosis
- Phagocytosis → Cell Eating
- Pinocytosis → Cell Drinking

Q) Give the differences between Passive diffusive and Active Transport

<u>PASSIVE DIFFUSION</u>	<u>ACTIVE TRANSPORT</u>
• Down Hill Transport	• Up Hill Transport
• No involvement of any Carriers	• Need of Carrier to transport the drug
• Down the Conc Gradient	• Against the Conc Gradient
• Equilibrium will be attained	• No equilibrium
• No expenditure of any Energy	• Utilization of Energy
• Not effected by Metabolic Poisons	• Effected by Metabolic Poisons
• No need of any O2	• Need of O2
• KE of the molecule guides the rate of transport	• Affinity towards enzyme & availability of energy guides the rate of transport
• Most of the drugs transport through this	• 5-FU via Pyramidine Transport • Niacin, Pyridoxine, Ascorbic Acid

Bioavailability

- The availability at the desired site and their work output
- Different brands of same generic drug can be *chemically equivalent*, may not be *therapeutically equivalent*
- *"the rate at which and the extent to which the active concentration of the drug is available at the desired site of action"* – **US FDA definition**
- **BIOAVAILABILITY** is an absolute term which requires measurement of both the true rate & total amount (extent) of drug that reaches the general circulation from an administered dosage form
- **EQUIVALENCE** is relative term which means comparison of one brand product with another of the same drug with a set of established standards.
- If two or more similar dosage forms of the same drug reach the blood circulation at the same relative rate and to the same relative extent → called bioequivalent of the generic drug.
- their plasma level profiles are comparable & super imposable
- **CLINICAL EQUIVALENCE:** The two brand products of one drug can be called as clinically equivalent if they provide an *"Identical in vivo pharmacological response "* as measured by control of symptoms or a disease.
- **THERAPEUTIC EQUIVALENCE** If one structurally different drug can provide the same therapeutic response or clinical response as another drug
- **CHEMICAL EQUIVALENCE:** If two or more dosage forms of the same drug contain the same labeled quantities of the drug as specified in Pharmacopoeia.
- C_{max}, t_{max} are simple indicators for the rate of absorption. An increased rate means a higher peak at a shorter time.
- C_{max} gives an indication whether the drug is sufficiently absorbed systematically to provide therapeutic response
- t_{max} gives an indication of the rate of absorption
- AUC, reflects the extent of absorption → it denotes the total amount of drug absorbed into the circulation during a specified period.

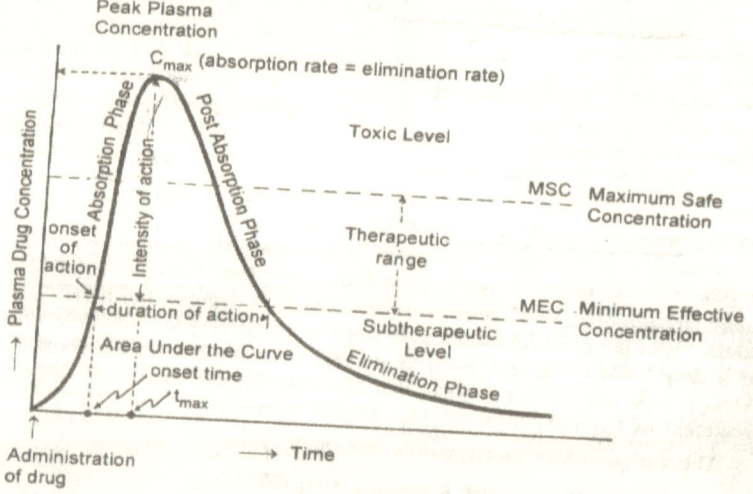

Factors affecting ABSORPTION & BIOAVAILABILITY

- *Drug Related & Patient Related Factors*

Drug Related Factors

- **Dosage Form** *(Solutions > Emulsions > > Suspension > Capsules > Tablets > Coated Tablet > Enteric > Sustained Release)*
- **Particle Size:** Smaller Particle Size
- **Salt Form:** Weakly acidic & Weakly basic drugs
- **Crystal Form:** Amorphous is better than Crystal
- **Water of Hydration:** Anhydrous forms are better than because molecules with water produces crystalline form
- **Degree of Ionization:** Non-ionized lipid soluble
- **Lipophilicity:** Unionized drugs if they are poor lipid solubility (low Ko/w) → poorly absorbed.
o Lipid solubility high enough to facilitate at absorption site & lipid solubility high enough to facilitate the partitioning of the drug in lipoidal bio-membrane & into systemic circulation
- **Drug pKa (Dissociation Const)**
▪ The lower the pKa of an acidic drug, stronger the acid i.e greater the proportion of ionised form at a particular pH
▪ The higher the pKa of a basic drug the stronger the base
- **For Acids**
1) *Very Weak acids (pKa>8)* such as Phenytoin, Ethosuximide & several barbiturates → unionized at all pH values & their absorption is rapid & independent of GI pH

2) *Acids pKa range 2.5 – 7.5* → greatly affected by changes in pH → absorption is pH dependent. Eg: Several NSAIDs → better absorbed in acidic condition (pH<pKa) → largely exists in unionised form.

3) *Stronger acids with pKa < 2.5* → Cromolyn Sodium → Poorly absorbed

- **For Basic drugs**
1) *Very Weak bases (pKa < 5.0)* such as Caffeine, Theophylline → unionized at all pH values → absorption is pH independent.

2) *Bases in pKa 5-11* → greatly affected Eg: Morphine analogues, Chloroquine, Imipramine

3) *Stronger Bases with pKa > 11.0* like Mecaamylamine & Guanethidine → are ionised → poorly absorbed

- **Product AGE & Storage conditions**

Patient Related Factors

AGE

- In infants, the gastric pH is high & intestinal surface & blood flow to the GIT is low resulting in altered absorption.
- In elderly, causes of impaired drug absorption include altered gastric emptying, decreased intestinal surface area & GI blood flow, higher incidents of achlorhydria & bacterial over growth in small intestine.

Gastric Emptying Time

- If gastric emptying time is less → peristaltic movements are fast → drug will not stay much time in the stomach → less absorption
- If gastric time is more → peristaltic movements are slow → drug will stay much times in the stomach → more absorption

FOOD

- *Food aids absorption* of Chloroquine, Carbamazepine, Griseofulvin, Spironolactone, Nitrofurantoin
- *Food delays absorption* of Ampicillin, Aspirin, Digoxin, Captopril, Levodopa, Penicillin G
- **Disease States   Drug-Drug Interactions   Blood Supply   1st Pass Effect   ROA**

DISTRIBUTION

- Pattern of scatter of the specified amount of drug among the various locations within the body.
- Drug is distributed to all organs including those not relevant to its therapeutic effect.
- Site of action, Protein binding, Tissue storage
- *Drug Disposition:* Distribution + Metabolism + Excretion → they decide the fate of the drug after absorption.
- it is a passive process, the driving force is concentration gradient between the blood & extra vascular tissues.
- It plays a significant role in the onset, intensity & duration of drug action.
- Differences in drug distribution among the tissues essentially arise as a result of a number of factors

REDISTRIBUTION

- Highly lipid soluble drugs given i.v. or by inhalation initially get distributed to organs with high blood flow e.g. brain, heart, kidney etc
- Later, vascular but more bulky tissues (muscle, fat) take up the drug – plasma conc falls & the drug is withdrawn from the sites.
- If the site action of the drug was in one of the highly perfused organs, redistribution results in termination of drug action.
- Greater lipid solubility of the drug, faster is its redistribution

Factors influencing Distribution

- Tissue Permeability of the drug
    - Physiochemical properties of the drug like molecular size, pKa & O/W partition coeff
    - Physiological Barriers
- Organ/Tissue size & perfusion rate
- Binding of drug to tissue components
    - Plasma Protein Binding
    - Binding to extra vascular tissue proteins
- Misc Factors
    - AGE   Diet   Obesity   Drug Interactions
    - Pregnancy   Disease States

Physiological Barriers to Drug Distribution

BLOOD BRAIN BARRIER

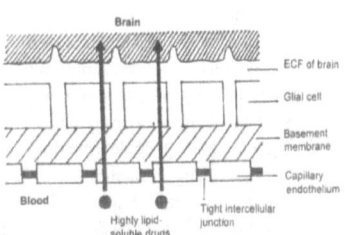

- Capillaries in the brain are highly specialized & much less permeable to water-soluble drugs.

- The brain capillaries consist of endothelial cells which are joined to one another by continuous tight intercellular

junctions comprising what is called as the Blood Brain Barrier.

- The presence of special cells called astrocytes, which are the elements of the supporting tissue fund a the base of endothelial membrane, form a solid envelope around the brain capillaries
- the molecules has to pass through them

- Drugs administered intranasally, diffuse directly into CNS due to continuity between submucosal areas of nose & subarachnoid space of the Olfactory lobe
- highly lipophilic drugs can only pass this

BLOOD PLACENTAL BARRIER

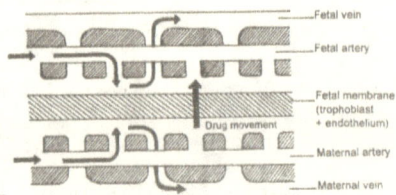

- The maternal & fetal blood vessels are separated by a number of tissue layer made of fetal trophoblast

basement membrane & the endothelium which together

constitute the Placental Barrier

- Many drugs having molecular weight less than 1000 D & moderate to high lipid solubility Eg: Ethanol,

Sulfonamides, Barbiturates, Gaseous Anesthetics, Steroids, Narcotic Analgesics & Anticonvulsants can diffuse very rapidly by simple diffusion.

- BPB is not as effective a barrier as BBB
- Nutrients essential for the fetal growth are transported by carrier-mediated processes.
- Immunoglobulins are transported by Endocytosis.
- Drugs are particularly dangerous to the Fetus during 2 Stages
- In the 1st trimester when the fetal organs, it is during this stage that most drugs show their Teratogenic effects (congenital defects) e.g. Thalidomide, Phenytoin, Isotretinion, Testosterone, Methotrexate.
- In the latter stages of pregnancy when drugs are known to affect physiologic functions e.g. Respiratory depression by Morphine.

V_d (Volume of Distribution)

- *Apparent V_d:* The hypothetical volume of body fluid into which a drug is dissolved or distributed.
- Total Body Water: Vascular Fluid/Blood (6 Lit) + Extracellular Fluid (Excluding Plasma) (12 Lit) + Intracellular Fluid (Excluding Blood Cells) (24 Lit) = 42 Lit
- If any drug is having Vd = 42 Lit means it is completely distributed through out the body.
- If any drug is having Vd < 42 Lit means it is confined to Systemic Circulation or Site of action or Protein Binding. Eg: Warfarrin
- If any drug is having Vd > 42 Lit means it is largely distributed & bind selectively to extra vascular tissues. → this kind of drugs leaves the body slowly & generally more toxic than drugs that do not distribute deeply into body tissues. Eg:Chloroquine
- Apparent Vd = Amount of drug in the body / Plasma Drug Conc
- Vd helps in estimation of the required dose of a drug $D_t = V_d (C_t-CO)$
- Vd also helps in correcting insufficient dose
- Vd also helps in the prediction of clinical consequences in displacement type of drug interactions.
- In diseases there will be alteration of V_d
- CHF, Uraemia, Cirrhosis of Liver → Alteration of V_d of Drug

PROTEIN BINDING

- The phenomena of complex formation with proteins is called as Protein Binding of Drugs.
- Protein Bind Drug is both Pharmacokinetically & Pharmacodynamically inert i.e. a protein bound drug is neither metabolized nor excreted nor is it pharmacologically active.
- Bound drug because of its enormous size, can't undergo membrane transport → half-life is increased.
- Binding involves weak Hydrogen bonds, Ionic bonds or van der Waal's forces → reversible.
- Binding of Drugs falls into 2 classes
  - Binding of Drugs to Blood Components
    - Plasma Proteins
    - Blood Cells
  - Binding of Drugs to Extravascular tissue proteins, fats, bones etc

PLASMA PROTEIN BINDING

- Drugs bind to Albumin, α1 acid glycoprotein, Lipoproteins, Globulins

- Highly plasma protein bound drugs remain largely restricted to vascular component

- Highly protein bund drugs ◊ very difficult to remove from the body

- Binding of drugs → Limited & Saturable Process
- Liver Disease, Uraemia → Hypoalbuminaemia → decrease in Binding → increase in Therapeutic Levels
- Physiological Stress, MI, Crohn's Disease, Inflammation → increase proteins
- **Displacement reactions**
  - Salicylates, Phenylbutazones, Valproic Acid, Naproxen, Sulfonamides → Displaces most of drugs
  - Phenylbutazone, Salicylates & Sulfonamides displaces Tolbutamide from its Binding Sites → Hypoglycaemia
  - Salicylates, Indomethacin, Phenylbutazone & Tolbutamide displaces → increases risk of haemorrhage
  - Sulfonamides and Vit K displaces Bilirubin → Kernicterus in Neonates
  - Salicylates displace Methotrexate
- Delays its Metabolic degradation & excretion
- Diminishes its penetration into CNS
- Assists drug absorption

Drug Reservoirs

> **Cellular Reservoir**
> ✓ When a drug has a higher affinity towards Cellular proteins than plasma proteins → then drug is primarily distributed in Tissues
> *Eg:*

1) Digoxin & Emetine in Skeletal Muscles, Heart, Liver & Kidney (Bound to Muscle proteins)
2) Iodine in Thyroid
3) Chloroquine in Liver (Tissue proteins)
4) Chloroquine in Retina (Nucleoproteins)

5) Cadmium, Lead & Mercury in Kidney (Muscle Protein, Metallothionein)
6) Chlorpromazine in Eye (Affinity to retinal pigment melanin)

➢ **Fat as Reservoir**
✓ Highly lipid soluble drugs, like Thiopentone, DDT & Phenoxybenzamine
✓ Sluggish reservoir due to lesser blood flow → if the body fats start depleting as occurs during starvation → the stored drug may be mobilized & toxicity may occur.

➢ **Transcellular Reservoir**
✓ Aqueous humor (e.g., Chloramphenicol & Prednisolone), CSF (e.g., Aminosugars & Sucrose), Pleural (e.g., Imipramine & Methadone), Pericardial & Peritoneal Sacs also serve as Drug Reservoirs

➢ **Bones & Connective Tissue**
✓ Tetracyclines, Cisplatin, Lead, Arsenic & Fluorides form a complex with Bone salts & get deposited in Nails, Bones, Teeth.
✓ Antifungal drug, Griseofulvin, has an affinity for keratin precursor cells & is selectively accumulated in the Skin & Finger nails
✓ Bone may become a reservoir for the slow release of toxic substances, like Lead, Arsenic & for Anticancer drug, like Cisplatin into Blood

➢ **Plasma Protein Binding**
✓ Albumin, α1 acid glycoprotein, Lipoproteins, Globulins

BIOTRANSFORMATION

- *Elimination:* The irreversible loss of drug from the body
- Elimination occurs by two processes viz. Biotransformation, Excretion
- *Biotransformation:* The conversion from one chemical form to another

3 different consequences with respect to Pharmacological activity of its Metabolite
- *Formation of an inactive metabolite from the active drug*
    - o   Eg: Pentobarbitone → Hydroxy Pentobarbitone
- *Formation of an active metabolite from an equally active drug*
    - o   Eg: Diazepam → Oxazepam, Codeine → Morphine
- *Formation of an active metabolite from an inactive (Prodrug) or a lesser active drug*
    - o   Eg: L-Dopa → Dopamine; Talampicillin → Ampicillin
- **PRODRUG** "Refers to a precursor drug that in itself has little or no Biological activity but is metabolized to a pharmacologically active Metabolite"
- An Ideal Prodrug should have following properties
1. Should not have intrinsic pharmacologic activity
2. It should rapidly transform, chemically or enzymatically, into the active form where desired
3. The metabolic fragments, apart from the active drug, should be non-toxic

Uses of Prodrugs
- *Pharmaceutical Applications*
    - ✓   Improvement of taste
    - ✓   Improvement of odor
    - ✓   Change of physical form for preparation of solid dosage forms
    - ✓   Reduction in GI irritation
    - ✓   Reduction of pain on injection
    - ✓   Enhancement of drug solubility and dissolution rate
    - ✓   Enhancement of chemical stability of drug
- *Pharmacokinetic Applications*
    - ✓   Enhancement of Bioavailability
    - ✓   Prevention of presystemic metabolism
    - ✓   Prolongation of duration of action
    - ✓   Reduction of toxicity
    - ✓   Site-specific drug delivery (drug targeting)

First Pass Metabolism

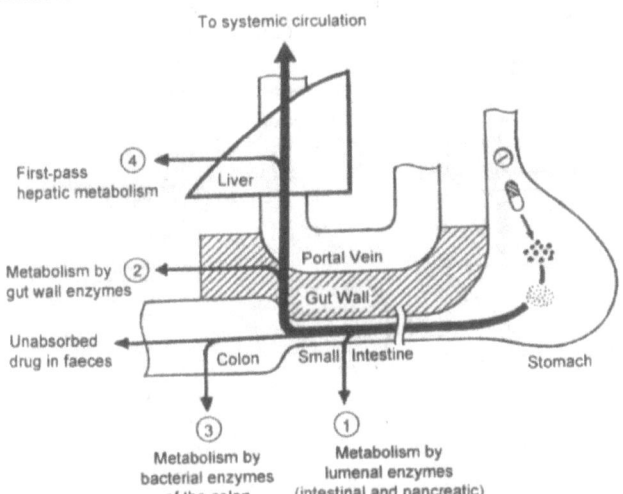

To systemic circulation

First-pass hepatic metabolism ④

Liver

Metabolism by ② gut wall enzymes

Portal Vein

Gut Wall

Unabsorbed drug in faeces

Colon | Small Intestine | Stomach

③ Metabolism by bacterial enzymes of the colon

① Metabolism by lumenal enzymes (intestinal and pancreatic)

- First Pass Metabolism or the Pre-Systemic Metabolism or the First Pass Effect → drug metabolism occurring before drug enters the systemic circulation.
- After Oral administration → drug crosses GIT wall → then through portal system before reaching systemic circulation → drug enters into Liver → metabolized by the CYP 450 family of enzymes.
- The net result is the decreased bioavailability of the drug & consequently a diminished therapeutic response → because of a significant amount of the drug is inactivated before reaching the systemic circulation.
- This can be bypassed by giving Parenterally or Sublingually
- In Liver diseases → bioavailability increases → Reduce the dose
- Exhibit shorter plasma $t_{1/2}$

BIOTRANSFORMATION
- Biotransformation Reactions are grouped into two types

PHASE – I

- These are degradative reactions
- The drug is diminished to a smaller polar/non-polar metabolite by introduction of a new group.
- mainly Microsomal, few are Non-Microsomal
- includes Oxidation, Reduction, or Hydrolysis

PHASE – II

- These are Synthetic Reactions or Conjugation Reactions
- catalyzed by Microsomal, Mitochondrial or Cytoplasmic enzymes
- The metabolite formed is usually Polar, Water Soluble & is most inactive

Drug Metabolizing Enzymes
- *Microsomal Enzymes*
- *Non-Microsomal Enzymes*

Microsomal Enzymes
- Associated primarily with Smooth surfaced Endoplasmic Reticulum of the Liver
- Principal enzymes → Mixed Function Oxidasess & CYP 450
- Microsomal enzymes are non-specific in action
- These are can be induced or activated & can metabolize only lipid soluble drugs
- These are concerned primarily with phase-I oxidation & reduction
- There is one unique microsomal enzyme system which carries out only one synthetic-phase-II reaction i.e. the glucuronic acid conjugation
- it is slightly different from almost all other microsomal enzymes in that
- it is the only microsomal enzyme system which is obtained in soluble cell fraction while other classical microsomal enzymes lose their activity in soluble form
- it is slightly different from almost all other microsomal enzymes in that
- it can form Glucuronide conjugates with a range of natural metabolites

Non - Microsomal Enzymes
- These are present in Cytoplasm, Mitochondria of hepatic cells & in Plasma
- These can be prepared as soluble fraction and still retain their catalytic property
- These include Monoamine Oxidase, Esterases, Amidases, Transferases & Conjugages
- Reactions catalysed by them are: All phase II reactions (except glucuronide conjugation), certain oxidations, reductions & hydrolytic reactions.
- These are non-inducible But show genetic variation

PHASE – I REACTIONS

Oxidations	Reductions	Hydrolysis
• Microsomal Oxidations	• Microsomal Reductions	• Microsomal Hydrolysis
• Non-Microsomal Oxidations	• Non-Microsomal Reductions	• Non-Microsomal Hydrolysis

Microsomal Oxidations

1) Aromatic Hydroxylation : *Phenobarbital → p-hydroxyxphenobarbital*
2) Aliphatic Hydroxylation : *Pentobarbital → Hydroxypentobarbital*
3) N – Dealkylation : *Morphine → Noemorphine*
4) O – Dealkylation : *Phenacetin → Paracetamol*
5) S – Dealkylation : *6-Methylthiopurine → Mercaptopurine*
6) N – Oxidation : *Trimethylamine → Trimethylamine N-oxide*
7) S – Oxidation : *Chlorpromazine to Chlorpromazine Sulfoxide*
8) Deamination : *Amphetamine → Phenylacetone*
9) Desulfurisation : *Parathion → Paraoxon*

Non – Microsomal Oxidations

1) Mitochondrial Oxidations : *Epinephrine → Vinyl-mendalic acid* by Monoamino Oxidase
2) Cytoplasmic Oxidations : *Alcohol → Acetaldehyde → Acetic Acid* by Alcohol dehydrogenase & Aldehyde dehydrogenase
3) Plasma Oxidative Processes : *Histamine → Imidazole acetic acid* by Histaminase

Microsomal Reductions

1) Nitro Reduction : *Chloramphenicol → Arylamine Metabolite*
2) Azo Reduction : *Prontosil → Sulfonamide*
3) Keto Reduction : *Cortisone → Hydrocortisone*

Non – Microsomal Reductions

1) *Chloral hydrate → Trichloroethanol*

Microsomal Hydrolysis

✓ Hydrolytic reactions are rarely microsomal, except for the hydrolysis of Pethidine to Pethidinic acid & Hydrolysis of Lidocaine by hepatic membrane bound esterase.

Non – Microsomal Hydrolysis

✓ Mainly for esters & amides
✓ Procaine to PABA by plasma choline esterase
✓ Atropine to Tropic acid

PHASE – II REACTIONS (Synthetic or Conjugation Reactions)

<u>Microsomal Conjugation</u>

I. <u>Glucuronide Conjugation</u>
   ❖ Parent drugs or their phase-I metabolites that contain Phenolic, Alcoholic, Carboxylic, Amino or Mercapto groups → undergo conjugation reaction with Uridine Diphosphate Glucuronic acid catalyzed by microsomal UDP-glucuronyl transferase enzymes → glucoride conjugates → highly polar → easily excreted
   ❖ Drug + UDPGA → Drug Glucuronide + UDP
   ❖ This is catalyzed by Glucuronyl Transferase (M)
   ❖ Eg: Morphine, Paracetamol, Aspirin, Chloramphenicol, Diazepam, Sulfonamides
   ❖ Bilirubin is also conjugated by Glucoronide
   ❖ Some high molecular weight conjugated metabolites of certain drugs are excreted in Bile.
   ❖ But bile is conjugated with Glucuronide
   ❖ Drug Glucuronide complex goes into Bile → Drug + Bile Glucuronide → Drug is again absorbed into systemic circulation → leads prolongation of the effect

<u>Non-Microsomal Conjugation</u>

II. <u>N-Acetyl Conjugation (in Cytosol)</u>
   ❖ This reaction is catalyzed by N-acetyltransferase which utilizes Acetyl CoA
   ❖ This is found in cell cytoplasm of various organs
   ❖ Eg: Aromatic Amines like Isoniazid, PAS, Dapsone, Sulfonamides, Procainamide and Histamine

III. <u>Sulfate Conjugation (in Cytosol)</u>
   ❖ This is catalysed by a family of enzymes called Sulfotransferases (present in cell cytoplasm of various organs)
   ❖ Cofactor – 3' Phosphoadenosine-5'-phosphosulfate (PAPS)
   ❖ These are highly polar compounds → rapidly excreted in urine
   ❖ Eg: Aspirin, Methyldopa, Paracetamol, Hydroxy-coumarins, Corticosteroids & Chloramphenicol

IV. <u>Amino Acid Conjugation (in Mitochondria)</u>
   ❖ Acyl-Coenzyme-A derivatives of carboxyllic acid drugs like Aspirin, Benzoic acid, Nicotinic acid & Deoxycholic acid → couple with Glycine or Glutamine in presence of AcCoA-glycine transferase enzymes in Miochondria
   ❖ Minor metabolic pathways

V. <u>Methyl Conjugation (in Cytosol)</u>
   ❖ Many Catecholamines like Dopamine & Epinephrine are O-Methylated while amines (like Histamine) are N-Methylated .
   ❖ The Catalyzing enzyme is transmethylase while cofactor (methyl donor) is S-Adenosine methionine.
   ❖ This is also a minor Metabolic pathway

VI. <u>Glutathione Conjugation (in Cytoplasm)</u>
   ❖ Many Epoxides, NO2 group containing drugs & Hydroxylamines like Ethacrynic acid & Sulfobromophthalein undergo glutathione conjugation.
   ❖ The catalyzing enzyme is Glutathione-S-transferase which is present in Cytosol or in Microsomoes
   ❖ This is also a minor pathway

VII. <u>Ribosides and Riboside Phosphate</u>
   ❖ Many Purines & Pyrimidines used as Antimetabolites in Cancer Chemotherapy form their active metabolites by forming ribonucleosides & ribonucleotides
   ❖ This is also a Minor Pathway

<u>FACTORS AFFECTING DRUG METABOLISM</u>

AGE: Neonates have low microsomal enzyme & glucuronyl transferase enzyme activity and they attain full function only few weeks after the birth
- They are unable to metabolize the given drug
- This occurs especially with Chloramphenicole → inability to metabolization leads to life threatening condition "Gray Baby Syndrome" (pg 675 KDT)
- The diminished activity of Glucuronyl Transferase in Neonates → Kernicterus
- Above 60 years → reduced hepatic blood flow → Drugs like Propranolol, Lidocaine & Pethidine (whose rate of Metabolism is governed by Hepatic blood flow) exhibit slow Metabolism & increased Toxicity

SEX: More important in animals than in Humans
- Male rats sleep for a shorter duration than females, after receiving hexobarbital
- Microsomal enzyme activity is higher in male rats

SPECIES: Rabbits metabolise Atropine faster than man as they have high atropine esterase activity in the liver and plasma

RACE: Most important
- Chinese have a high alcohol dehydrogenase but a low aldehyde dehydrogenase activity (Higher conc of Aldehyde → Headache, Palpitation)
- Chinese are more sensitive to CV effects of Propranolol, Metabolize faster than Whites

Nutrition & Diet: The diet rich in Protein & low in carbohydrate enhances the rate of metabolism of many drugs
- Rich protein diet supplies Glycine & Cysteine which are essential to form conjugated metabolites of the drug
- Deficiency of Vitamin A, Vitamin C, Calcium & Magnesium also impair the drug metabolism & increase the individual's sensitivity to different drugs.

DISEASE: Viral Hepatitis, Alcoholic Hepatitis, active or inactive Liver Cirrhosis, Hepatocellualr Carcinoma & Heavy metal poisoning → Impairs the activity of CYP 450
- Hypothyroidism decreases the metabolism of Digoxin, Methimazole & Practolol.

Genetic Variation: Some genetic variation leads to decrease in Metabolism or increase in Metabolism

Genetic defect	Drug	Therapeutic Use	Clinical Consequences
Atypical Pseudocholinesterase (inherited as autosomal recessive trait)	Succinyl Choline	Neuromuscular Blocker	Prolonged Aponea due to retarded metabolism
Slow N-acetylation	Isoniazid	AntiTB	Peripheral Neuropathy
Rapid N-acetylation	Isoniazid	AntiTB	Hepatotoxicity
Faulty expression of CYP-450-2D6	Debrisoquin	AntiHTN	Orthostatic hypotension
Faulty expression of CYP-450-2D6	Bufuralol	Beta – Blocker	Enhanced beta blockade & nausea
Faulty expression of CYP-450-2D6	Sparteine	Antiarrhythmic	Oxytocic symptoms

- **Drug-Drug Interactions:** Same as thing as in Enzyme Induction and Enzyme Inhibition

ENZYMATIC INDUCTION

- Several drugs, on repeated administration, stimulate or induce the growth of smooth endoplasmic reticulum
- This leads to an enhanced microsomal enzyme activity. But it is a slower.
- It results in an accelerated metabolism and consequently in a decreased pharmacological response.
- Most prominent in Liver, can also be in Lung, Placenta & Kidney eg CYP-450-1A1 isoenzyme is induced in the lungs of cigarette smokers.

Clinical Significance of E Induction

- Unwanted pregnancy → in the users of Oral Contraceptive Pills if potent enzyme inducers like Phenytoin or Rifampicin are used.
- Patients on enzyme inducing drugs like Barniturates, would need higher doses of Oral Anticoagulants like Warafarrin → if the Baribiturates are suddenly withdrawn then the dose of Warfarrin also should be reduced → otherwise patient will may start bleeding.
- Enzyme Inducers like Phenytoin → accelerate the metabolism of Vit D3 → leads to Osteomalacia
- Enzyme Inducers like Barbiturates enhance their own metabolism leading to Metabolic Tolerance
- Enzyme Induction leads to Toxicity eg: Ethanol drinkers have more probabilities of developing Hepato-toxicity from Paracetamol
- even in therapeutic doses also due to increased production of N-acetyl-p-benzoquinone, a hepato toxic metabolite of Paracetamol
- can also be useful in therapeutic benefits eg: to treat neonatal jaundice
- can also be useful in therapeutic benefits eg: to treat neonatal jaundice
- Phenobarbitone can be given to the pregnant mother, 7-14 days prior to labour or to the infant soon after the birth to induce the foetal hepatic glucuronyl trasnferase enzyme which catalyses the conjugation of bilirubin to glucuronic acid → metabolized and excreted rapidly

ENZYMATIC INHIBITION

- One drug may inhibit the Metabolism of another drug due to the inhibition of microsomal enzymes → increase in the levels in plasma → increase in therapeutic levels or increase in toxicity
- it is a rapid process and reversible

Potentially Adverse Consequences

- Unexpected nausea/vomiting due to Theophylline with concomitantly administered Chloramphenicol
- Enhanced bleeding tendency with Dicumarol when administered with Cimitidine
- Severe respiratory depression with morphine when given with MAOIs (Mono-Amine Oxidase Inhibitors)
- Severe Ataxia & Drowsiness with phenytoin in combination with dicumarol or Chloramphenicol
- Precipitation of Cardiac arrhythmias with Tefenadine when given in combination with Chloramphenicol

Therapeutically Beneficial Consequences

- Increased accessibility of L-Dopa in brain when given along with Carbidopa (L-aminoacid decarboxylase inhibitor)

- Aversion to alcohol after prior administration of Disulfiram (Aldehyde dehydrogenase inhibitor)
- Reversal of skeletal muscle paralysis due to d-Tubocurarine by Neostigmine (Acetylcholinesterase reversible inhibitor)

Excretion of drugs
Billiary Excretion & Enterohepatic circulation
- Drugs that show significant elimination through bile include: Quinine, colchicine, vinblastine, d-tubocurarine, vecuronium, corticosteroids, erythromycin & chlorpromazine → these drugs subsequently excreted through feces
- Some drugs are secreted through bile, but after being delivered to intestine, are reabsorbed back and the cycle is repeated (enterohepatic circulation) → e.g., Digitoxin
- Certain drug metabolites (particularly the glucuronides) are excreted through bile & delivered to the intestine where these metabolites are deconjugated or hydrolysed releasing the parent active drug again → this free drug is then reabsorbed & the cycle is repeated (enterohepatic circulation) → e.g., Thyroxine, Morphine, Chloramphenicol, Tetracycline, Phenolphthalein, Ethinyl estradiol
- The net result of enterohepatic circulation is the prolongation of drug action as the drug serves as a small circulating reservoir

Fecal Elimination
- Mostly orally ingested drugs are excreted through faeces
- E.g., magnesium sulfate, streptomycin, neomycin, bacitracin, cholestyramine
- Certain drugs which are excreted through bile but not reabsorbed from intestinal tract
- E.g., erythromycin, chlorpromazine, corticosteroids

Alveolar Excretion
- General anesthetics, nitrous oxide, ether, paraldehyde & Alcohol

Excretion through Breast Milk
- Drugs excrete through milk will show pharmacological effects in nursing infant
- Drugs are transferred to the breast milk according to pH partition principle
- Basic drugs predominantly unionized at plasma alkaline pH → can diffuse through mammary epithelium → get accumulated in the milk → can't be reabsorbed back to plasma → at relatively acidic pH of milk they get ionized → excreted in milk
- E.g., Chloramphenicol, tetracyclines, morohine, metronidazole, carbimazole, oral contraceptive pills, diazepam, antihistamines
- Certain acidic drugs → being less secreted → cause serious side effects in infants
- E.g., Sulfonamides (Kernicterus & Allergy), Penicillins (Allergy), Ampicillin (Diarrhoea), Dapsone (Haemolytic Anemia), Phenindione (Bleeding), Phenobarbitone (Drowsiness), Phenytoin (methaemoglobinaemia), Theophyline (restlessness)

Excretion through skin, Hair, Sweat & Saliva
- Griseofulvin is secreted through keratin precursor cells
- Arsenic, Mercury salts & iodides → secreted through hair follicles
- Iodine, Potassium iodide, lithium & Phenytoin → secreted through Saliva
- Certain Amines & urea → secreted through sweat

First Order Kinetics	Zero Order Kinetics
• A constant fraction of the drug is eliminated at a constant interval of time	• A constant or a fixed quantity of the drug is eliminated (or absorbed) per unit time
• The rate of elimination is directly proportional to the plasma concentration	• The rate of elimination is proceeding at a fixed rate, independent of the concentration of the drug in the plasma
• If the plasma concentration is increased → rate of elimination also increases proportionately	• Increasing the dose doesn't result in a proportionate rise in the extent of elimination

• T1/2 always remain constant irrespective of the dose	• T1/2 of the drug is never constant
• If Log plasma concentrations are plotted against time → plasma fall out curve is LINEAR	• If Log plasma concentrations are plotted against time → plasma fall out curve is CURVILINEAR
• Drug elimination never becomes 100%	• Drug elimination becomes zero
• If the dose of the drug is doubled → duration of action is prolonged for one more half life	• No change
• Drugs having high Vd will have two half-lives values (distribution t1/2, elimination t1/2)	

Clearane: The theoretical volume of plasma from which drug is completely removed in unit time

Alteration of Urinary pH

• Acidic drugs are more ionized in Alkaline urine → so they are removed from the body very easily
• Basic drugs are more ionized in Acidic urine → so they are removed from the body very easily
• Normal pH of the Urine is slightly basic → alteration of the pH of the urine can leads either prolongation of the drug action or faster elimination of the drugs
• Drug action can be prolonged just by altering the pH of urine which is especially useful where drugs are excreting fastly → Acidic drugs will show more prolonged drug action in Acidic urine & Basic drugs will show more prolonged drug action in Alkaline urine
• Drugs can be excreted very rapidly just by altering the pH of the urine which is especially useful in renal impairment and drug overdose toxicity → Acidic drugs can be quickly removed by altering the urine into Alkaline & Basic drugs can be quickly removed by altering the urine into Acidic
• In Salicylate and Barbiturate overdose poisoning cases →alkalinizing the urine makes them to excrete rapidly as they are acidic drugs and they ionize rapidly in alkaline urine
• In Morphine and Amphetamine overdose poisoning cases → Acidifying the urine makes them to excrete rapidly as they are basic drugs and they ionize rapidly in acidic urine

Pharmacodynamics

• Study of the biochemical & physiological effects of drugs & their mode of action
• What a drug does to the body

The Site of Drug Action
• The drug may act at extra-cellular, cellular or intracellular sites

Extra-cellular Site of Action
• Antacids neutralizing gastric acidity
• Chelating agents forming complexes with heavy metals

Cellular Site of Action
• Action of Acetylcholine on nicotinic receptors of motor end plate
• Effect of sympathomimetics on heart muscle & blood vessels

Intracellular Site of Action
• Sulfa drugs act by interfering with the synthesis of folic acid which is intracellular component

Mechanism of Action of Drugs
• Drugs act either by receptor mediated or by non-receptor mediated mechanisms or by targeting specific genetic changes

Receptor Mediated Mechanisms
• Most of the drugs produce their effects by interacting with the receptors

- Binding of drug with receptor results drug-receptor complex → which is responsible for triggering the biological response
- **Affinity:** the capability of a drug to form the complex with its receptor
- **Efficacy or Intrinsic Activity:** the ability of a drug to trigger the pharmacological response after making the drug-receptor complex
- **Agonists:** which have both the high affinity as well as high intrinsic activity & can trigger the maximal biological response or mimic the effects of endogenous substance after combining with the receptor
- **Antagonists:** which have only the affinity but no intrinsic activity & can block or interfere with the binding of an endogenous agonist
- **Partial Agonists:** these have full affinity to the receptor but with low intrinsic activity & these are only partly as effective as agonists
- **Inverse Agonists (Negative Antagonists):** these have full affinity towards the receptor, but their intrinsic activity ranges between "zero to minus one"

Receptor Types & Signal Transduction Mechanisms

- Four receptor families

Type – I

- Agonist regulated ion channels (known as ligand gated ion channels)
- Localized on cell membrane
- The flow of ions through these channels can elicit cellular response in the form of depolarization or hyperpolarisation of the cell membrane
- E.g. Nicotinic – Cholinergic receptor, GABA receptor, Glutamate receptor, Glycine receptor

Type – II

- Membrane bound receptors → coupled to effector system through GTP binding proteins called as G-proteins
- E.g. Muscarinic – Cholinergic receptors, Adrenoceptors, Dopaminergic receptors, 5-HT receptor, Opiate receptor, Purine receptors
- There are 3 G-Protein Coupled Effector Systems
  o Adenylate Cyclase: Cyclic AMP system: drugs produce their effects by increasing or decreasing the catalytic activity of adenylate cyclase → raising or lowering the concentration of cAMP within the cell
    ▪ E.g. regulation by cAMP dependent protein kinase include the increased activity of $Ca+2$ channels in the heart muscle
  o Phospholipase – C: Inositol Phosphate System: drugs bind to receptors linked to G-proteins & activate the membrane enzyme phospholipase – C (PLC) → leads to hydrolysis of phosphotidyl inositol 4,5-diphosphate → splits into two second messengers: Diacylglycerol (DAG) & Inositol 1,4,5-tri-phosphate (IP3) → IP3 triggers the release of $Ca+2$ from storage vesicles → released $Ca+2$ binds to calmodulin & Calmodulin – $Ca+2$ regulated the activity of various enzymes
    ▪ DAG activates a phospholipids & the calcium sensitive protein kinase-C → response
  o Ion channel regulation: GPCR control the functioning of ion channels (e.g. K+ & $Ca+2$ channels)
    ▪ Don't involve any second messengers

Type – III

- Directly linked to tyrosine kinase (e.g. receptors for insulin & various growth factors) or to guanylate cyclase (e.g. receptors for atrial natriuretic peptide)
- These are the receptors that are ligand (agonist) activated transmembrane enzymes having catalytic activity

- The receptor is autophosphorylated → activates the receptor tyrosine kinase activity towards intracellular signaling proteins → response

Type – IV

- Cytosolic receptors for steroid hormones, thyroid hormones & Vit D
- The agonist, first penetrates the cell to combine with the steroid binding site of the receptor → binding then unfolds the receptor & exposes the normally masked DNA-binding site → then associates with DNA of specific genes → stimulates the RNA polymerase activity & specified m-RNA is synthesized → m-RNA then directs the production of specific proteins → shows the response

Receptor Desensitization

- Receptor mediated responses often 'desensitise' with the time
- After reaching an initial high level, the response gradually diminishes over seconds or minutes even in the continuing presence of the agonist
- It is usually reversible
- At the NM junction, the desensitization is caused by a slow conformational change in the receptor resulting in the tight binding of the agonist molecule without opening of the ion channel

Up & Down regulation of Receptors

- Prolonged exposure to high concentration of agonist causes a reduction in the number of receptors available for activation → down regulation of receptors → this is due to Endocytosis or internalization of the receptors from the cell surface
- On prolonged use of Salbutamol in severe asthmatic patients is due to down regulation of β_2 receptors
- Endogenous depression → down regulation of alpha receptors
- Prolonged use of tricyclic antidepressants results down regulation of beta receptors
- Prolonged occupation of receptors by a blocker leads to an increase in the number of receptors with subsequent increase in receptor sensitivity → up regulation of receptors → this is due to externalization of the receptors out again from inside of the surface
- E.g., in thyrotoxicosis, the thyroid hormones bring about up-regulation of β_1 receptors of cardiac muscle which increases cardiac sensitivity of catecholamines leading to tachycardia

Non-Receptor Mediated Mechanisms

By Chemical Action

- Neutralization: Antacids act by neutralizing gastric hyperacidity
- Chelation: some drugs (chelating agents) trap the heavy metals in their ring structure & form water soluble complexes which are then finally excreted e.g. EDTA, Desferroxamine
- Ion Exchangers: anion exchange resin like cholysteramine exchanges Cl. ions from the bile salts → the resultant complex is not absorbed & is excreted out

By Physical Action

Osmosis   Adsorption   Protectives   Demulcents
                Astringents

The Qualitative Aspects of Drug Effects

Adverse Drug Reactions

- Mainly two types: **Type – A ADRs (expected undesirable effects), Type – B ADRs (unexpected undesirable effects)**
Type – A ADRs (expected undesirable effects)

<u>Side Effects</u>

- These are undesirable effects which are observed even with the therapeutic doses of the drug & are mainly mild and manageable
- E.g., Dicyclomine relieves pain of intestinal colic due to its antispasmodic action but side-by-side also causes dryness of mouth even in its therapeutic doses
- Promethazine has antiallergic action & also produces sedation in therapeutic doses
<u>Secondary Effects</u>

- These are indirect consequences of the main pharmacodynamic action of the drug
- E.g., development of super infection after suppression of bacterial flora by antibiotics & weakening of host defenses after the use of corticosteroids
<u>Toxicity</u>

- These are exaggerated form of side effects which occur predictably either due to overdoses or afer prolonged use of the drug
- The reason can be pharmacodynamic or pharmacokinetic
- E.g., bleeding due to high doses of heparin or coma due to high doses of barbiturates
- Crystaluria or glomerular nephritis due to precipitation of sulfonamides in acidic urine
- nephrotoxicity due to gentamicin in cases having renal insufficiency
- delirium, hyperpyrexia & hallucinations with overdoses of atropine
3 therapeutic strategies are used to avoid toxicity

- The drug should always administered at the lowest dose that produces acceptable benefit
- Adjunctive drugs that act through different receptor mechanism
  - Prazosin → decreases BP in essential hypertension by acting → as alpha1 antagonist in vascular smooth muscle → but patient → postural hypotension (increased dose) → concurrent administration of diuretics & vasodilators → decreased dose of Prazosin → no side effects
- Selectivity of the drug's actions may be increased by manipulating the concentration of the drug available to receptors in different parts of the body
  - Anatomic selectivity may be achieved → by changing the route of drug administration
Type – B ADRs (unexpected undesirable effects)

- These arise unexpectedly, even when the drug is used in therapeutic doses, by a mechanism unrelated to the main pharmacological effect of the drug

Drug Allergy (Hypersensitivity Reactions)

- Allergic responses to drug occur when there has been previous exposure to drug (or its metabolites) & when this sensitized individual is re-exposed to the same drug again

- In the begining stage → drug acts as Hapten which after combining with host proteins → antigenic

- Specific antibodies are formed against his antigen → these antibodies keep circulating

- On re-exposure → antigen – antibody response → release of chemical mediators of allergy [5-HT, Leukotrienes, SRS – A, PAF] → effects like urticaria, rhinitis, pruritis, asthma, anaphylactic shock

- Immediate → hypotensive shock, bronchoconstriction, angioneurotic, laryngospasm → death

Types of Allergy

Type – 1 (Immediate Type)	Type – 2 (Acute or Accelerated)	Type – 3 (Delayed)	Type – 4 (Cell Mediated)
■ Develops within mins & lasts for 2 – 3 hrs ■ Formation of tissues sensitizing IgE → fixed to mast cells or leucocytes ■ Exposure to drug → degranulates mast cells → activates leucocytes → release of chemical mediators of Allergy → anaphylactic shock	■ Occurs within 72 hrs of drug administration ■ Formation of IgG & IgM ■ On re-exposure → activates compliment system → fever, lupus erythematous, haemolysis, thrombocytopenia	■ Occurs within 1 – 2 weeks of drug administration ■ Formation of IgG → deposited on vascular endothelium → activates compliment ■ Allergic inflammatory reactions in tissues, glomerular nephritis & serum sickness	■ Antigen specific receptors → developed on T-lymphocytes ■ Local or tissue allergic reactions (photo sensitization & contact dermatitis)

Cross Allergy

- Within the members of same group of drugs is most common

- Genetically Determined Abnormal Responses of a Drug

- Mechanisms underlying such unusual responses to be of Genetic origin

- Variations due to single mutant gene (genetic polymorphism)

Pharmacogenetic Variations in Phase – 1 drug metabolism

- **Presence of atypical pseudo choline esterase (faulty hydrolysis):** Inheritance of this trait is autosomal recessive
  - Neuromuscular blocking action of Succinyl Choline is terminated within 5 min by hydrolysis due to normal pseudocholine esterase in plasma

- Genotype atypical pseudocholine esterase can't hydrolyse succinyl choline (may take 1-2 hr) → prolonged respiratory failure
- Atypical pseudocholine esterase can easily be detected by measuring its inhibition by dibucaine
- **Hydroxylase polymorphism (faulty oxidation)**
  - Phenytoin is hydroxylated & thus oxidized by mixed function oxidases
  - In slow hydroxylators → Phenytoin toxicity increases
  - Defect is transmitted as autosomal recessive trait
  - Other drugs → Tolbutamide, Phenformin, Nifedipine

Pharmacogenetic Variations in Phase – 2 Drug Metabolism

- **Acetylator Status:** Many drugs are metabolized by hepatic N-acetylase enzyme (not inducible)
  - Variations in drug response not due to inducers
  - Variations are due to the presence of higher or lower amounts of N-acetylase in the liver
  - *Rapid Acetylators:* Eskimos & Japanese
  - *Slow Acetylators:* Egyptians, Mediterranean, Swedes
  - Acetylator status of an individual significantly affects the nature of ADRs with drugs which are mainly metabolized through N-acetylation
  - Examples: Isoniazid, Procainamide, Hydralazine, Dapsone
  - In Slow Acetylators: Isoniazid gets accumulated after repeated doses → neurotoxicity → because Isoniazid inhibits pyridoxine kinase which converts pyridoxine to its active form pyridoxyl phosphate
  - Addition of Vit B6 controls the side effects
  - In Fast Acetylators: Isoniazid gets metabolized faster → accumulation of acetyl hydrazine → hepatotoxicity
  - Slow acetylation of Procainamide or Hydralazine may cause antinuclear antibodies to develop in plasma → systemic in Systemic Lupus Erythematous
  - Dapsone, in slow acetylators → haemolysis

Pharmacogenetic Variation in Drug Response due to Enzyme Deficiency

- **G6PD Deficiency:** in RBCs is inherited as a sex linked recessive trait
  - It is common in Africans & American Negros
  - G6PD responsible for supply of NADPH which acts as a cofactor for glutathione reductase (enzyme which converts oxidized form of glutathione to its reduced form)
  - Reduced form of glutathione converts Ferric (Fe^{+3}) to Ferrous (Fe^{+2}) in haemoglobin
  - Reduced glutathione important for stability to RBC membrane prevent their haemolysis
  - Drugs with oxidizing property like Primaquine, Sulfonamides, Sulfones, Nitrofuranoin cause haemolytic anaemia in G^PD deficiency
- **Uroporphyrinogen Synthetase** → for haem synthesis → barbiturates, Carbamazepin, Phenytoin, Chloramphenicol, Oral Contraceptives → precipitating attacks of intermittent porphyria

Idiosyncratic Drug Responses

- Condition of malignant hyperpyresia → due to halothane, succinyl choline, chlorpromazine, haloperidol → as inherited trait

- Aplastic anaemia → due to Chloramphenicol → Idiosyncratic effect due to presence of p-NO2 in Chloramphenicol
- Aspirin induced late onset asthma or chronic renal failure
- Thiazide diuretics induced erectile impotence

Teratogenicity

- A teratogen is an agent which can cause a physical fetal malformation when maternal administration the period of organogenesis
- After the period of organogenesis, during the pregnancy, exposure to fetotoxic drugs may cause functional disability or an alteration in growth of the organs/fetus but no physical defects
- E.g., the sedative 'thalidomide' prescribed to pregnant women for giving relief from morning sickness, was found to produce various types of developmental anomalies in the newborns → the commonest anomalies were 'amelia' or total absence of one or more limbs, & 'phocomelia' or absence of one or more limbs

Iatrogenic Disease

- Drugs themselves may produce certain pathological syndromes
- Such diseases produced as a result of therapeutic measures are known as 'Iatrogenic diseases'
- These are also known as physician's disease
- E.g., Glucocorticoid therapy can precipitate hypertension, CHF & Cushing's syndrome
- Glucocorticoids, Aspirin & Indmethacin may precipitate perforation of duodenal ulcer

The Quantitative Aspects of Drug Effects

- **Dose:** it is the required amount of drug in weight, volume, moles or international units, that is necessary to provide a desired effect
- In clinical practice → it is known as Therapeutic dose
- In experimental works → it is known as Effective dose
- **Total Dose:** it is the maximum quantity of the drug that is needed during the complete course of the therapy
- **Loading Dose:** it is the large dose of the drug to be given initially to provide the effective plasma concentration rapidly
- **Maintenance dose:** it is half of the loading dose given after the loading dose to maintain the steady state plasma concentration

Therapeutic Index (Margin of safety): a measure of the relative desirability of a drug for the attaining of a particular medical end that is usually expressed as the ratio of the largest dose producing no toxic symptoms to the smallest dose routinely producing cures

- o Therapeutic Index = LD_{50} / ED_{50}
- o The larger the therapeutic index, the safer the drug
- o For safe therapeutic application of the compound, its therapeutic index must be more than one

Therapeutic Window: the range of dosage of a drug or of its concentration in a bodily system that provides safe effective therapy

Factors influencing Drug Action

- **Age, Body, Weight & Body Surface Area**
    - *Yonug's Formula*

        Child's Dose = (Age in years / (Age +12)) X Adult Dose

    - *Dilling's Formula*

        Child's Dose = (Age in years / 20) X Adult Dose

    - *Clark's Formula*

        Child's Dose = (Wt of child (kg) / 70) X Adult Dose

    - *According to BSA*

        Child's Dose = (BSA (m^2) / 1.8) X Adult Dose

        BSA (m^2) = W$^{0.425}$ (kg) X H$^{0.725}$ (cm) X 0.000718

- **Sex:** the differences in drug disposition during menstruation, pregnancy & lactation may necessitate am alteration of drugs for women
- **Environment & Time of Drug Administration**
    - Slightly higher doses of sedative-hypnotics are needed to induce sleep in day light than at night
    - At high altitudes, the capacity of the body to oxidize drugs (phase – I metabolism) is diminished → therapeutic doses produce toxicity
- **Psychological & Emotional Factors**
    - Nervous & anxious patients require more dose of general anesthetic and of anxiolytic drugs
- **Genetic Factors & Idiosyncrasies**
    - Modified responses due to genetic variations in group of people (refer in above text)
- **Metabolic Disturbances & Pathological State**
    - Disease induced abnormality may also modify a drug effect
    - Low acidity decreases iron absorption → results in a decreased response
    - In celiac disease, the absorption of amoxicillin is decreased while that of cephalexin & co-trimaxazole is increased
- **The Route & Frequency of Drug Administration**
    - The route of administration governs the speed & intensity of drug response

Drug Tolerance

- Tolerance is characterized by the need to increase the dose in order to produce the pharmacological response of equal magnitude & duration

- The log dose – response curve of a tolerant person shows a shift towards right side because higher than initial doses are required to achieve the same effect

- Tolerance is a common phenomenon seen with CNS active drugs, like morphine, alcohol, barbiturates, amphetamine

Drug Disposition or Metabolic Tolerance

- This type of tolerance may occur when a drug reduces its own absorption or increases its own metabolism through microsomal enzyme induction → the decrease in the effective concentration of the drug at the site of action
- The excess doses taken by the tolerant person, because the excess doses taken by the tolerant person are simply compensating for the losses incurred due to faster drug disposition
- E.g., chronic alcoholics can tolerate large amounts of alcohol because of their thickened (indurated) gastric mucosa which reduces the extent of alcohol absorption

Cellular Adaptive (target tissue or pharmacodynamic) Tolerance

o This type of tolerance may be attributed to some kind of adaptive changes that have taken place within the system after repeated drug administration
o This might be due to 1) drug induced changes in the receptor density (down-regulation) 2) impairment in receptor coupling to signal transduction pathways
o The log dose response curve shifts towards right → because , after repetitive dosing, although the plasma concentrations are rising, yet there is no proportionate rise in the drug response due to cellular adaptation of the biological system
o E.g., morphine, caffeine, nicotine

Acute Tolerance (Tachyphylaxis)

▪ The acute development of tolerance after a rapid & repeated administration of a drug at shorter intervals

▪ Gradual depletion of the agonist from the storage sites with no chances of its replenishment because of the repeated administration of the drug at short intervals

▪ E.g., tachyphylaxis see with sympathomimetic drugs like ephedrine, amphetamine, tyramine

▪ Tachyphylaxis may also occur as a result of a change in the sensitivity of target cells

▪ E.g., tachyphylaxis to nitroglycerine is observed among workers exposed to this drug while in its manufacturing industry

Drug Dependence

▪ Repeated administration of certain drugs may induce a habit & dependence

▪ If the drug is not available after dependence → he develops withdrawal symptoms like headache, restlessness & emotional upset, convulsions & vasomotor collapse

▪ *WHO definition of* **Drug Dependence** " A state, psychic & sometimes also physical, resulting from the interaction between a living organism & a drug, characterized by behavioral & other responses that always include a compulsion to take the drug on a continuous or periodic basis in order to experience its psychic effects & sometimes to avoid the discomfort of it absence. Tolerance may or may not be present. A person may be dependent on more than one drug"

▪ A condition in which a drug produces "a feeling of satisfaction and a psychic drive that require periodic or continuous administration of the drug to produce pleasure or to avoid discomfort" is called Psychic Dependence

▪ It is characterized by a craving for the drug & an overwhelming concern for obtaining and using it

▪ In Physical Dependence, the body "achieves" an adaptive state that manifests itself by intense physical disturbances when the drug is withdrawn

▪ Drug addiction or drug abuse is used o denote the phenomenon involving both psychic & physical dependence on drugs

▪ Its characteristics are

✓ An overpowering desire to taking the drug

✓ A tendency to increase the dose

✓ A high tendency to withdrawal symptoms

▪ Drugs causes Psychic Dependence: Amphetamine, Nalorphine

■ Drugs cause Psychic & Physical Dependence: Codeine, Heroin. Ethyl Alcohol, Nicotine (Tobacco)

Photosensitivity

■ It is a cutaneous reaction resulting from drug induced sensitization of the skin to UV radiation

■ The reactions are of two types

Phototoxic

■ Drug or its metabolite accumulates in the skin, absorbs light & undergoes a photochemical reaction followed by a photobiological reaction reslting in local tissue damage (sunburn lilke) i.e. erythema, edema, blistering followed by hyperpigmentation & desquamation

■ E.g., Tetracyclines

Photoallergic

■ Drug or its metabolite induces a cell mediated immune response which on exposure to light of longer wave lengths produces a popular or eczematous contact dermatitis like picture

■ E.g., Sulfonamides, Sulfonylureas, Griseofulvin

Combined Effects of Drugs

■ **Summation:** when two drugs elicit the same response, but with different mechanism, & their combined effect is equal to the algebraic sum of their individual effects, the drugs are said to exhibit summation of effects

■ **Additive Effects:** combined effect of two drugs, acting by the same mechanism, is equal to that expected by simple addition

Synergism: when the combined effect of the two drugs is greater than the algebraic sum of their individual effects, the phenomenon is called as synergism

● The net come is either the potentiation or prolongation of effects

● This results when two drugs act at different sites or when one drug alters the pharmacokinetics of the other drug

● E.g., Sulfamethoxazole combined with trimethprim (acting at two different sites); Levodopa + carbidopa (change of pharmacokinetic profile)

● The total amount of dose can be reduced and also less side effects will be there

Antagonism: any time when the conjoint effect of two drugs is less than the sum of the effects of the individual drugs → phenomenon is known as Antagonism

There FOUR types of Mechanisms exists

● Chemical Antagonism: when the two drugs are chemical antidotes to each other
    ✓ Heparin is antagonized by protamine is a chemical antagonism

● Physiological or Functional Antagonism: when two agonists, acting at different sites, counterbalance each other by producing opposite effects on the same physiological system
    ✓ CNS stimulates antagonize the action of CNS depressants

33

- Biological or Pharmacokinetic Antagonism: when one drug affects the absorption, metabolism or excretion of the other & effectively reduces the concentration of the active drug at its site of action
  - ✓ Urinary alkalisers like Sod. Bicarbonate & Sod. Dihydrogen citrate increase the urinary excretion of acidic drugs
- Pharmacological Antagonism: it is a pharmacodynamic antagonism between two drugs action on the same receptor

➢ **Reversible (or competitive or equilibrium type) Antagonism**
- Agonist and the antagonist bind to the same receptor and there will be competition for binding between them
- Response of antagonist can be overcome by increasing the concentration of agonist
- Log dose response curve of agonists shifts towards right in presence of antagonist and the efficacy will not change
- ED_{50} will be higher in presence of antagonist
- The duration of competitive blockade is short due to higher rate of dissociation of the antagonist molecule from its receptor sites
- E.g., atropine is a competitive antagonist of acetylcholine

➢ **Irreversible (non-equilibrium type) Antagonism**
- The dissociation of antagonist fro the receptor will be very less when agonist concentration is increased → no change in the response of antagonist
- Antagonist bind to receptor through covalent bond → agonist unable to replace antagonist
- Higher agonist concentrations can't overcome the response of antagonists → maximal agonist response can't be obtained
- E.g., Phenoxybenzamine, an irreversible alpha blocker, or Methysergide, an irreversible 5-HT blocker

➢ **Non-Competitive Antagonism**
- The antagonism shows its action through another receptor or through other mechanisms
- There will not be any competition between the agonist and the antagonist
- Antagonist is not chemically resembles with the agonist
- There will be change in both efficacy and potency of the agonist in presence of antagonist
- Log dose response curve flattens after maximal response (due to unsurmountable antagonism)
- The response is entirely dependent on the concentration of antagonist
- E.g., Diazepam - Bucuculline

➢ **Negative Antagonism (Inverse agonism)**
- These antagonists stabilize the receptor from undergoing the productive agonist-independent conformational change & produce that are opposite to those of agonist
- E.g., β – carbolines are negative antagonist at benzodiazepine receptors

Drug Combinations

Drug – Drug Interactions

- An alteration in the effectiveness or toxicity of one drug due to the action of another simultaneously administered drug is known as drug – drug interaction
- Drug interaction may result in
  - ✓ Adverse effects wherein there is a decrease in the effectiveness or an increase in the toxicity of one drug in the presence of the other
  - ✓ Biological interference with laboratory tests which may mislead the diagnosis

✓ Beneficial effects, wherein there is an increase in the effectiveness or a decrease in the toxicity of one drug in the presence of another

➢ **Pharmacokinetic reason include:**

● *Alteration in absorption:* liquid paraffin decreases the absorption of fat soluble vitamins A, D, E

● *Alteration in distribution:* Salicylates displace Tolbutamide from protein binding and causes Hypoglycemia & Sulfonamides displace Warfarrin from protein binding and causes Haemorrhage

● *Alteration in metabolism:* enzyme inducers and enzyme inhibitors

● *Alteration in excretion:* urine alkalizers increases excretion of acidic drugs

➢ **Pharmacodynamic reason include**

● Different types of Antagonism: refer above text for examples

➢ **Drug interaction which interfere with diagnostic tests:** Nalidixic acid, Salicylate & Vit C provide a false positive test of urine sugar with Benedict's solution

➢ **Drug interactions lead to beneficial effects**

● In therapeutics (examples of synergism)

● In the management of poisoning (antagonism)

Fixed Dose Combinations

- the combination of two different drugs in a single pharmaceutical formulation
- E.g., Co-Trimaxazole (combantion of Sulfamethoxazole & Trimethprim), Combination of Amoxicillin & Clavulanic acid

Advantages
- Convenience in dose schedule & better patient compliance
- Enhanced effect of the combination
- Minimization of side effects

Disadvantages
- The dose of any component drug can't be adjusted independently if desired
- If the pharmacokinetic profile do not match → leads to fluctuations in steady state concentrations
- Becomes difficulty to identify which drug is causing the adverse drug reactions

Essential Drug List Concept

- Essential Drugs are those **drugs *that satisfy health care needs of the majority of population; these should therefore be available at all times, in adequate amounts and in appropriate dosage forms***
- WHO published its first model of Essential Drug List (EDL) in 1977
- It is updated for every 2-3 years in view of changing world scenario about the medical needs & availability of new & better drugs
- WHO has recently connotated a new term called "Essential Medicines" in place of Essential Drugs
- Essential Medicines are defined as those that *satisfy the priority health care needs of the population. They are selected with due regard to public health relevance. Medicines are intended to be available within the context of functioning health systems at all times in adequate amounts, in appropriate dosage forms with assured quality & adequate information & at a price the individual & the community can afford* [14th model of Essential Medicined List in 2005]
- WHO's EML is not feasible to any particular country because those medicines may not be essential to a particular country
- Eg: New Zealand is snake free country → so anti-snake venom is not essential

- But Indian subcontinent & African countries are full of snakes → anti-snake venom is essential
- The selection of medicines for EML by any country should be an ongoing process by taking into account the changing health priorities and epidemiological conditions as well as advancement in the knowledge of medicines

Criteria for Selection of Essential Medicines

- To meet the needs of the majority of people, priority should be given to medicines of proven efficacy and safety.
- Unnecessary duplication of medicines and dosage forms should be avoided.
- Newly released similar medicines should only be included if they have distinct advantages over similar products currently in use
- Each selected medicine must meet the prescribed standard of quality, including bioavailability and shelf life in locally prevailing climatic and storage conditions
- When two or more medicines appear to have similar efficacy and safety, the preference should be given to that medicine which

(a) has been most thoroughly investigated,

(b) has most favorable pharmacokinetic data &

(c) can be reliably manufactured with indigenous resources, i.e., which could be made available locally in good quality as well as quantity and at a lower cost

- The cost of the total treatment including the cost of any investigation if needed thereafter (to assure the safety of the medicine) should be considered as major selection criterion for inclusion of any medicine in EML
- Cost:Benefit Ratio of the medicine or of its dosage form should serve as the criterion for its inclusion in the EML of any country
- The proposed medicine, should preferably be a single drug compound. FDC of two medicines is acceptable only when these two medicines meet the requisite pharmacokinetic criteria for their combination (half lives & V_d)
- FDC have advantage over the individual medicine as regards therapeutic effect, safety, patient compliance & the cost
- The medicines included in EML should bear the International Non-proprietary Names

Advantages of EDL/EML

- It makes for wider availability of medicines to most strata of the population, particularly meagre and limited resources have to be utilized effectively
- It helps in making substantial savings in terms of foreign exchange without affecting the basic health care needs
- It helps improvement in drug dispensing and patient compliance, by substantially curtailing the number of drugs available in hospitals
- It helps planers and policy makers to allocate reasonable funds and resources for supply of medicines.
- Even procurement and distribution of medicines can be made systematic and uniform at all levels of health care system
- The advanced medical care institutes & teaching hospitals can even compile their own formularies based on this list and depending upon the specialised facilities available at their centres.

Rationale Use of Drugs (RUD)

• The Rational Use of Drugs (RUD) is based on the Rule of Right: that means the *right drug*, given to the *right patient*, in *right dosage*, at the *right cost*
• It also fulfil the SANE criteria, which means Safety, Affordability, Need and Efficacy of the drug should always be considered first before prescribing it to the patient
• Rational prescribing, involves a right decision of the prescribing doctor for the benefit of the patient
• Rationale Use of Drugs (RUD)
• Acording to WHO, RUD requires that *"patient should receive medication appropriate to their clinical needs, in appropriate doses, for an appropriate period at the lowest cost"*

IRRATIONAL PRESCRIBING

TYPES: Incorrect-, Inappropriate-, Over-, Under-, Multiple-prescribing

• *Over-prescribing → leads to many undesired effects, leads to overdose, inconvenient for the patient to take so many drugs, valuable & scarce resources are wasted*
• *Under-prescribing → treatment is not effective, aggressive or an expensive treatment may be needed later*
• *Incorrect → wrong treatment, adverse effects, situation may become worse*
• IRRATIONAL PRESCRIBING
• *Inappropriate → needs for longer durations use of drugs may leads to adverse effects, more hospital stay, inconvenient to patient, psychological stress*
• *Multi-prescribing → giving multiple prescriptions*

Concept of P-Drug

• How to choose the right drug for each patient to achieve the set therapeutic goal as per the diagnosis of the disease?
• A drug selected by the doctor is called as "P-Drug"
• This "P" may stand for Preferred or Particular drug chosen by the to treat a disease effectively at a reasonable low cost (i.e., in a cost –effective manner)

How to select P-Drug
• P-drug is never the one that has been suggested/dictated to you by your senior colleague or by a sales representative
• *Firstly* because the latest or the most expensive drug may not be the best, safest and the most cost effective
• *Secondly* because your patient's well being is your responsibility which you can't pass on to others if the patient is not benefited by your treatment
Steps in choosing a P-Drug

1) **Define the Diagnosis:** be specific in the diagnosis rather in general [e.g., rather writing angina pectoris, better to diagnosis as Stable angina pectoris which is caused by partial occlusion of the coronary arteries]

2) **Specify the therapeutic Objective:** Patient is having acute attack of angina → immediate requirement is to provide quick relief from pain → this can be achieved by increasing the oxygen supply to or by reducing the oxygen demand of the cardiac muscle

Steps in choosing a P-Drug

Therapeutic objective in this situation:

(a) by decreasing the pre-load,

(b) by decreasing the contractility,

(c) by decreasing the heart rate or

(d) by decreasing the after –load of the cardiac muscle

3) Make a list of Effective Group of Drugs: there are 3 groups of drugs with such an effect: The Nitrates, the Beta blockers & the Calcium channel blockers

4) Choose an Effective Group by Comparing Efficacy, Safety, Suitability & Cost of Treatment

- *Efficacy:* Objective is that the drug should work as soon as possible
- *Nitrates* → cause peripheral vasodilatation & decrease oxygen demand by decreasing pre-load & after-load of the cardiac muscle
- *Calcium channel blockers & Beta blockers* → decrease pre-load, contractility, heart rate & after load
- All the groups contain drugs or their dosage forms with a rapid effect.
- Oral Nitrates → have high first pass metabolism → sublingual is used
- Calcium channel blockers and Beta blockers → injectable forms are available
- *Safety:* All the groups side effects are more or less
- *Suitability:* For a patient of angina pectoris → dosage form of the drug should be such that it can be administered by himself during the attack & which should guarantee a rapid effect → nitrates have edge over other drugs
- *Cost of the Treatment:* Nitrates are usually inexpensive drugs as compared to other group of drugs
- hence Nitrates are the chosen group of drugs

5) Choose P-Drug from the effective group of drugs

- 3 drugs are available: Glyceryl trinitrate, isosorbide mononitrate, isosorbidedinitrate
- among them Glyceryl trinitrate is the cheaper one

Superinfection (Suprainfection)
- Appearance of a new infection as a result of antimicrobial therapy
- AMAs cause some alteration in the normal microbial flora of the body
- The normal flora contributes to host defence by elaborating substances called "Bacteriocins" → which inhibit pathogenic organisms
- The more "Broad" the effect of an antibiotic on microorganisms, the greater the alteration in the normal microflora and greater the possibility that a single microorganism will become predominant invade the host, and produce infection
- The incidence of Superinfection is lowest with Penicillin G
- Higher with Tetracyclines and Chloramphenicol
- Highest with combinations of broad spectrum antimicrobials and the expanded spectrum 3rd generation Cephalosporins

- Further development & use of Broad spectrum agents will lead to more extensive alterations in the normal flora → more super infections
- Superinfections are more common when host defence is compromised, as in
- Corticosteroid therapy
- Lukemias & other malignancies anticancer drugs → immunosuppresants and decrease WBC count]
- Acquired immuno deficiency syndrome
- Agranulocysis
- Diabetes, diseminated lupus erythematous
- Sites involved in Superinfection are those that normally harbour commensals i.e., Oropharynx, Intestinal, Resp, Genitourinary Tracts, Ocassionally skin
- Superinfections are generally more difficult to treat.
- The organisms frequently involved, manifestations and drugs for treating Superinfections are
  o Candida albicans: Monilial Diarrhoea, Thrush, Vulvovaginitis → treat with Nystatin or Clotrimazole
  o Resistant staphylococci: Enteritis → treat with Cloxacillin
  o Proteus: UTI, Enteritis → treat with a Cephalsporin or Gentamicin
  o Clostridium difficile: Pseudomembranous entercolitis associated with the use of Clindamycin, Tetracyclines, Aminoglycosides, Ampicillin, Cotrimoxazole → most common after Colorectal Surgery → metrinidazole & Vancomycin are DOC
  o Pseudomonas: UTI. Enteritis → treat wih Carbenicillin, Piperacillin, Gentamicin

To minimize Superinfections
- Use specific (narrow spectrum) AMA whenever possible
- Don't use antimicrobial to treat trivial self limiting or untreatable (VIRAL) infections
- Don't unnecessarily prolong antimicrobial therapy
- **Minimum Inhibitory Concentration:** The lowest conc of an Antibiotic which prevents visible growth of a bacterium
- **Minimum Bactericidal Concentration:** The concentration of Antibiotic which kills 99.9% of bacteria

Autonomic Nervous System

Autonomic Nervous System (Cholinergic)

Autonomic Nervous System consists of main 3 divisions

- ✓ Parasympathetic Nervous System (or Cholinergic)
- ✓ Sympathetic Nervous System (or Adrenergic)
- ✓ Enteric Nervous System

Parasympathetic Nervous System

- Acetylcholine is the neurotransmitter at all autonomic ganglionic synapse
- Hypothalamus is the major controlling centre for the parasympathetic system
- Acetylcholine Receptors are Muscarinic and Nictotinic receptors

[Basic Mechanism of any receptors

Stimulation (Excitatory) → Muscle Contraction, Glands Secretion ↑, relaxation of Sphincters

Inhibition → Muscle relaxation, Glands Secretion ↓, contraction of Sphincters]

Muscarinic Receptors

- The receptors are located at the para symp neuroeffector junction at all smooth muscles & glands
- These are primarily divided into 5 subtypes (M_1, M_2, M_3 are most important)

M_1 Receptors (Neuronal & Gastric)

- *Location:* Ganglia (autonomic & enteric), Gastric, Paracrine cells, CNS (cortex & hippocampus)
- *Function:* gastric acid secretion, GI motility, CNS excitation
- *MOA:* ↑ IP_3, ↑ DAG, Cytoplasmic Ca^{+2}, Depolarization

M_2 Receptors (Cardiac)

- *Location:* SA node, AV node, Atrium, Ventricle, Presynaptic terminals
- *Function:* SA node: ↓rate of impulse generation; AV node: ↓velocity of conduction, ↓contractility, vagal bradycardia
- *MOA:* Inhibition of adenylate cyclase (↓cAMP) & opening of K^+ channels; Inhibits neuronal Ca^{+2} channels (Presynaptic inhibition of Ach release)

M_3 Receptors (Glandular)

- *Location:* Exocrine glands, smooth muscles, vascular endothelium
- *Function:* ↑exocrine secretions, smooth muscle contraction

41

- *MOA:* ↑ IP$_3$, ↑ DAG, Cytoplasmic Ca^{+2}, Depolarization

Nicotinic Receptors

- These receptors are located in neuromuscular junctions & at all autonomic ganglia
- They play a facilitatory role in the release of other transmitters like Dopamine & Glutamate
- They are classified as Muscle type (N$_M$), Neuronal type (N$_N$) & Central nicotinic receptors

"Cholinomimetics" means the drugs which imitate or mimic the actions of Ach at the neuroeffector junction and also called as *"Parasympathomimetics"*

Acetylcholine

- ACh is of no therapeutic value → due to its ultra short action (very rapidly hydrolyzed by the enzyme acetylcholinesterase present in the synaptic cleft) & secondly because of it widespread activity at all cholinergic sites throughout the body
Biosynthesis:

- ACh is synthesized in the axon terminal from Choline & Acetylcoenzyme-A by the cytosolic enzyme choline acetyltransferase
Metabolism

- ACh is hydrolyzed by an enzyme called Acetylcholinesterase to Choline & Acetic acid.
- Choline is again utilized for the biosynthesis of ACh

Acetylcholinesterase enzyme is of two types

1. *True ACh Esterase:* it is a membrane bound enzyme present in the cholinergic synaptic cleft
○ It is very highly specific in hydrolyzing ACh & other acetylesters
○ It is mainly localized in neuronal membrane, cholinergic synaptic cleft & to a small extent in RBS & placenta
2. *Plasma Choline Esterase (pseudo choline esterase or butyryl choline esterase):* it is synthesized in liver & found predominantly in plasma & in intestine.
○ Its actions are non-specific as it hydrolyses succinyl choline, benzoylcholine & butyrylcholine esters more easily than ACh

ACTIONS

Eye

- Circular muscle of iris, ciliary muscle & lacrimal glands posses M$_3$ receptors
- Radial muscle of iris & eylid smooth muscles have no para symp supply
- ACh causes contraction of circular muscle of iris & ciliary muscle

- Contraction of circular muscle of iris → causes Miosis → also opens the pores of the canal of Schlemm (aided by stretching of pupil due to miosis) → which facilitates drainage of aqueous humor → better drainage reduces IOP
- Contraction of ciliary muscle → makes the suspensory ligaments of the lens loose → makes the lens more convex (by reducing its focal length) → eye's focus is accommodated for near vision
- Stimulation of M_3 receptors at lacrimal glands → produces lacrimation (due to vasodilatation)

Salivary Glands

- Salivary glands → M_3 → stimulation produces watery saliva → due to vasodilatation resulting from the release of bradykinin

Lungs

- Smooth muscle of bronchi → M_3 → stimulation causes bronchoconstriction & mucous glands → M_3 → stimulation causes more bronchial secretions

Gastro Intestinal Tract

- GIT smooth muscle → M_3 → stimulation causes increase in the motility & increase tone of the GUT smooth muscle
- Sphincters → M_3 → stimulation causes relaxation of sphincters
- Gastric glands → M_3 → stimulation causes increased secretions
- Gastric parietal cells → M_1 → stimulation promotes gastric acid secretions

Heart

- Para symp supply is only up to SA node, Atria & AV node but not below
- Ventricular myocardium has Muscarinic receptors but no innervation → can exhibit a decrease in contractile strength (only to exogenously administered ACh)
- SA node → M_2 → stimulation causes a decrease in the heart rate (negative chronotropy) & decrease in the contractile strength (negative inotropy)
- AV node → M_2 → stimulation causes a decrease in the conduction velocity & increase in the refractory period

Blood Vessels

- Arteries have no para symp innervation, but have M_3 → exogeneous cholinomimeitc cause fall in BP due to vasodilatation
- The endothelium of most blood vessels release EDRF (endothelium relaxing factors like NO & other cotransmitters) → causes vasodilatation due to M_3 activation by exogenous cholinomimetics.
- Systemic veins have neither para symp innervation nor M-receptors

Urinary Bladder

- Detrusor muscle → M_3 → stimulation causes contraction
- Sphincter → M_3 → stimulation causes relaxations
- Both these actions lead to voiding of urinary bladder

Pancreas

- Acni cells → M_3 → stimulation causes increased secretion of pancreatic juice

Male Sex Organs

- Vascular bed of erectile tissue is dilated while venous sphincters are closed leading to erection of penis (might be due to stimulation of M_3)

Sweat Glands

- Sweat glands → M_3 → stimulation causes increased sweating

Central Nervous System

- Tremors & Convulsions → due to Muscarinic effects
- Ataxia, behavioural disturbances & restlessness → due to Nicotinic effects

Classification (Parasymapathomimetics or Cholinomimetics)

➢ **Directly Acting**
1. Acetylcholine
2. Synthetic Choline Esters
- Methacholine, Carbachol, Bethanechol
3. Natural Alkaloids
- Muscarine, Nicotine, Pilocarpine, Arecoline
4. Misc
- Tremorine, Oxotremorine
➢ **Indirectly Acting (Anticholinesterases)**
❖ *Reversible (short to intermediate duration of action)*
1. Natural Alkaloids : Physostigmine
2. Quaternary Compounds: Edrophonium, Neostigmine, Pyridostigmine, Ambenonium, Demecarium, Rivastigmine
❖ *Irreversible (long duration of action)*
1. Organophosphates : Isoflurophate (DFP), Ecothiophate, Paraoxon, Parathion, Malathion, Diazinon
2. Carbamates: Propoxur

Muscarine

- It is obtained from poisonous mushrooms *Amantia muscaria*
- It crosses BBB very little because it is quaternary compound
- **Symptoms** of Mushrooms poisoning are Miosis, Sweating, Salivation, Diarrhea, increased micturation, decrease in HR (all the actions of ACh on Muscarinic receptors)
- Treatment: Parenteral administration of Atropine 1-2 mg i.m. every 30 min till symptoms subside with adequate supportive measures for respiration, circulation & pulmonary oedema
- Mushroom poisoning is also called as *Mycetism*

Pilocarpine

- It is the chief alkaloid obtained from the leaves of the shrub *Pilocarpus jaborandi*
- It is a tertiary amine so it rapidly crosses BBB
- It has main M_3 action and mild N_N actions
- Pilocarpine is too toxic for systemic use as it produces usual effects of choline esters and pulmonary oedema
- Main therapeutic uses include
- Ophthalmic Use:
a) For the initial treatment of open angle glaucoma where it is instilled into eye as 0.5% - 0.4% solution
- Reduction in IOP occurs within few mins & lasts for 4 – 8 hrs
b) To counteract the mydriasis produced by Atropine
c) To break adhesions between the iris and the lens (as in Iridocyclitis) where it is instilled alternatively with homatropine (a mydriatic)
- As Sialagogue: Rarely, Pilocarpine (5 – 10 mg orally) is used to stimulate salivary secretions in patients after laryngeal surgery

Q) Give the Pharmacological basis for the use of Pilocarpine in Glaucoma

Ans: Pilocarpine has main M_3 action → reduces IOP

➢ Circular muscle of iris, ciliary muscle & lacrimal glands posses M_3 receptors
➢ ACh causes contraction of circular muscle of iris & ciliary muscle
➢ Contraction of circular muscle of iris → causes Miosis → also opens the pores of the canal of Schlemm (aided by stretching of pupil due to miosis) → which facilitates drainage of aqueous humor → better drainage reduces IOP
➢ Contraction of ciliary muscle → makes the suspensory ligaments of the lens loose → makes the lens more convex (by reducing its focal length) → eye's focus is accommodated for near vision
➢ Stimulation of M_3 receptors at lacrimal glands → produces lacrimation (due to vasodilatation)
➢ For the initial treatment of open angle glaucoma where it is instilled into eye as 0.5% - 0.4% solution
- Reduction in IOP occurs within few mins & lasts for 4 – 8 hrs

Anticholinesterases or AChE inhibitors or Indirectly acting Parasymapthomimetics

▪ These drugs inhibit acetylcholinesterase (AChE) enzyme, which is present in synaptic cleft & responsible for rapid hydrolysis of ACh → so these drugs prolong the action and increase the availability of ACh at the Muscarinic and/or nicotinic receptors after its release from postganglionic parasympathetic neurons.

Reversible (Competitive) Inhibitors of AChE

▪ Includes: **Physostigmine, Neostigmine, Pyridostigmine, Edrphonium, Ambenonium, Demecarium & Rivastigmine**
▪ The reversible anticholinesterase drugs bear a structural resemblance to AChE → they combine with the anionic & esteratic sites of AChE → this complex is less readily hydrolysed than AChE-ACh complex → it results in a temporary inhibition of the enzyme (because AChE regeneration takes longer time) → prolongs the duration of action of ACh released in synaptic cleft
▪ Among these Edrophonium has a shorter duration of action (10min) → because, Edrophonium forms complex only at the anionic site → thereby reversibly preventing the binding of ACh with AChE
▪ Physostigmine is a tertiary amine → more lipid soluble → can cross BBB easily → centrally acting
▪ Remaing drugs are quaternary amines → less lipid soluble → can't cross BBB easily → peripheral actions

Physotigmine

▪ It's a tertiary amine
▪ It is highly lipid soluble & shows better absorption in all body components including CNS
▪ It acts on M_1, M_2, M_3
▪ It also stimulates ganglia
▪ Being highly toxic → only limited use
Therapeutic Uses

▪ Ophthalmic use
a) To counteract the effects of mydriatics after refraction testing
b) To prevent adhesions between iris & the lens or iris & cornea resulting due to irutis, iridocyclitis & corneal ulcer
c) For the treatment of Glaucoma
▪ In the treatment of Atropine poisoning (Belladona poisoning): Physostigmine is a specific antidote for atropine poisoning or for poisoning by any other anticholinergic drug
● Being a tertiary amine it can cross BBB easily & can antagonize central as well as peripheral toxicity
● Initially, the diagnosis is made by giving smaller doses of Physostigmine (0.5 to 1.0 mg i.m.) → if normal parasympathomimetic effects of Physostigmine (flushing, sweating, salivation, lacrimation) are not observed → it could be a case of atropine poisoning
● Treatment: Physostigmine 2mg i.v. or i.m. initially or additional doses thereafter if necessary

Myasthenia Gravis

- It is an acquired autoimmune disorder causing skeletal muscle fatiguability & weakness
- It is associated with the production of IgG antibody that binds to ACh receptors (N_M) at the post-junctional motor end plate
- The reduction in the number of receptors (N_M) results in the reduction of the amplitude of the end plate potential which in turn fails to trigger an action potential
- **Symptoms:** weakness of the muscle & fatigue which worsens after the exercise but goes off after the rest, ptosis, diplopia, slurring of speech, difficulty in swallowing & weakness if extremities
- **Treatment:** Treatment is started with any reversible anti AChE agent of intermediate duration of action; the main drugs which are available (any one from the following)
Neostigmine 15 – 30 mg 6 hrly orally;

Pyridostigmine 60 – 120 mg, 4 – 6 hrly orally; or

Ambenonium 5 – 25 mg 6 hrly orally

- Pyridostigmine 60 -120 mg 4 – 6 hrly orally & Prednisolone 10 mg OD or an alternate day, to be increased slowly to a maximum of 100 mg OD or alternate day
- **Basis:** These drugs inhibit acetylcholinesterase (AChE) enzyme, which is present in synaptic cleft & responsible for rapid hydrolysis of ACh → so these drugs prolong the action and increase the availability of ACh at nicotinic receptors after its release from postganglionic parasympathetic neurons → more availability of N_M receptor

Paralytic Ileus

- ileus resulting from failure of peristalsis
- obstruction of the bowel ; specifically : a condition that is commonly marked by a painful distended abdomen, vomiting of dark or fecal matter, toxemia, and dehydration and that results when the intestinal contents back up because peristalsis fails although the lumen is not occluded
- a twisting of the intestine upon itself that causes obstruction
- **Treatment:** Neostigmine 0.5 – 1mg s.c.
- Basis: The reversible anticholinesterase drug bear a structural resemblance to AChE → it combines with the anionic & esteratic sites of AChE → this complex is less readily hydrolysed than AChE-ACh complex → it results in a temporary inhibition of the enzyme (because AChE regeneration takes longer time) → prolongs the duration of action of ACh released in synaptic cleft
- Due to the prolongation of ACh action, acts on M_3 receptors in GIT
o GIT smooth muscle → M_3 → stimulation causes increase in the motility & increase tone of the GUT smooth muscle
o Sphincters → M_3 → stimulation causes relaxation of sphincters

Atony of Urinary Bladder

- Loss of tone in urinary bladder
- Treatment: Neostigmine 0.5 – 1 mg s.c.
- Basis: The reversible anticholinesterase drug bear a structural resemblance to AChE → it combines with the anionic & esteratic sites of AChE → this complex is less readily hydrolysed

than AChE-ACh complex → it results in a temporary inhibition of the enzyme (because AChE regeneration takes longer time) → prolongs the duration of action of ACh released in synaptic cleft
- Due to prolongation of duration of action of ACh causes the following actions
○ Detrusor muscle → M_3 → stimulation causes contraction
○ Sphincter → M_3 → stimulation causes relaxations
○ Both these actions lead to voiding of urinary bladder

Post-operative Decurarization (Treatment of curare poisoning)

- Treatment: Neostigmine (0.5 - 1 mg i.v.) or Edrophonium (10 mg i.v.) along with Atropine → rapidly reverses muscle paralysis induced by d-tubocurarine given during anesthesia or by poisoning due to snake venom neurotoxin
- Prior atropinisation is more beneficial because it not only counteracts Muscarinic side effects of Neostigmine and also avoids the transient summation of bradycardia by these two drugs (atropine by blocking Presynaptic M_1 receptors on postganglionic parasymapathetic vagal nerve endings causes initial bradycardia)

Alzheimer's Disease

- Alzheimer's disease (AD) is the most prevalent form of dementia
- Marked decrease in choline acetyltransferase & loss of cholinergic neurons in brain → which account for much of the learning & memory deficit in AD
- Other neurotransmitter loss includes that of brain glutamate, dopamine, 5-HT & somatostatin
- Cholinesterase inhibitors like Rivastigmine, Tacrine, Donepezil, Galantamine are used in the treatment of AD
- These drugs block the degeneration of ACh & increase the availability of ACh in synaptic clefts
- Tacrine, a longer acting reversible anticholinesterase, can be used for palliative treatment of mild to moderate form of AD
- It is orally active & provides improvements inmemory, cognition
- Tacrine facilitates the release of ACh from cholinergic nerve endings
- Rivastigmine, Donepezil, Galantamine are newer anticholinesterases having better penetration into CNS
- These are less toxic and better tolerated

Irreversible inhibitors of AChE

- Organophosphorus compounds: **Diflos (DFP), Ecothiophate, Parathion, Malathion, Diazinon**
- Carbamate derivatives: **Proposure, Carbaryl**
- These drugs are pentavalent containing a labile fluoride group or a labile organic group
- These drugs are irreversible blockers because they phosphorylate the esteratic site of AChE irreversible by forming a covalent bond → during this, the labile group is released leaving the remaining part of the drug molecule attached covalently with the esteratic site of AChE through its phosphorus atom → AChE becomes inactive & complex becomes very stable (resistant to hydrolysis) due to covalent bonding)
- The phosphorylated AChE enzyme undergo a rapid process of "ageing"
- Ageing means that within a period of 1 – 2 hrs, AChE – drug complex undergoes molecular rearrangement & becomes completely resistant to hydrolysis (i.e. no reactivation of AChE)
- The ageing is due to the loss of one alkyl or one alkoxy group, leaving a much more stable monoalkyl or monoalkoxyl-phosphoryl-AChE complex

Pharmacological Effects (Symptoms of poisoning)

▪ These drugs are volatile non-polar substances of very high lipid solubility → they are rapidly absorbed through mucous membranes & unbroken skin
▪ Symptoms are manifested as combination of Muscarinic, Nicotinic and CNS side effects
1. *Muscarinic toxic manifestations:* Diarrhea, urination, miosis, bronchoconstriction, lacrimation, salivation, sweating, bradycardia & hypotension
2. *Nicotinic toxic manifestations:* Fasciculations of skeletal muscles leading to paralysis
3. *CNS toxic manifestations:* Restlessness, tremors, convulsions, ataxia & respiratory arrest

Treatment of Organophosphorus Poisoning

▪ Decontamination measures such as thorough cleaning of skin, giving bath using copious amount of water & soap with special attention to hair, nail & eyes. Remove contaminated clothes
▪ Atropine sulfate 2 mg i.v is the mainstay of treatment. It is repeated every 15 min until full atropinization occurs i.e dilation of pupils take place & all the Muscarinic symptoms & signs are reversed
▪ Atropine is then given as maintenance dose at 12 hourly interval depending upon severity condition
▪ Pralidoxime (2 PAM) which is a cholinesterase reactivator should be given within 24 – 48 hr (if it is not given promptly aging of enzyme occurs & reactivation does not occur after poisoning
▪ The dose is 1 – 2 gm dissolved in 250 ml of 5% glucose solution & injected over 15 – 30 min
▪ It relieves the nicotinic effects (muscle weakness, twitching & respiratory depression & also Muscarinic effects of organophosphorus poisoning
▪ Pralidoxime is hazardous in poisoning by carbamate insecticides
▪ Other measures: keep the airway clear, start mechanically assisted pulmonary ventilation, give gastric lavage with activated charcoal
▪ Lastly observe the patient closely for at least 72 hrs with constant monitoring of cardiac & pulmonary functions

Manifestations of Ageing

▪ Delayed neurotoxicity in the form of severe polyneuritis, ataxia, reduced tendon reflex, weakness & ultimately flaccid paralysis
▪ No specific therapy is known
▪ Pralidoxime has most marked action at skeletal neuromuscular junction & almost insignificant effect in CNS & autonomic effector sites

Basis of Pralidoxime in Organophosphorus poisoning

1. Organophosphoruscompounds phosphorylate the esteratic site of AChE irreversible by forming a covalent bond → during this, the labile group is released leaving the remaining part of the drug molecule attached covalently with the esteratic site of AChE through its phosphorus atom → AChE becomes inactive & complex becomes very stable
2. Pralidoxime reactivates the enzyme AChE by attaching with the anionic site which lies vacant in the phosphorylated enzyme

3. Oxime group in the Pralidoxime is closer to phosphorylated esteratic site, attracts phosphate group (phosphate transfer to –NOH group of 2-PAM) → the oxime-phosphate complex diffuses out, leaving the regenerated AChE enzyme in an active form

- All oximes are ineffective as antidotes if poisoning has occurred due to carbamate group of anti-AChE drugs (e.g., Propoxur)
- Carbamates attach themselves with anionic site is not free for attachment to Pralidoxime which is prerequisite for their mode of action

Antimuscarinic Drugs or Anticholinergics

Anticholinergic

- Anticholinergic includes both Antimuscarinic and Antinicotinic drugs

Classification of Antimuscarinic Drugs

1. Natural alkaloids
➤ Atropine (dl-hyoscyamine)
➤ Scopolamine (l-hyoscine)

2. Semi synthetic derivatives: Homatropine, Atropine methionitrate, Hyoscine methylbromide, Anisotropine, Benatropine, Ipratropium bromide, Tiotropium bromide

3. Synthetic derivatives
a) Tertiary amines: Eucatropine, Cyclopentolate, Tropicamide, Dicyclomine, Flavoxate, Oxybutinin, Pirenzepine, Telenzepine, Trihexyphenidyl, Procyclidine, Biperiden

b) Quaternary amines: Propantheline, Methantheline, Oxyphenonium, Glycopyrrolate, Clidinium, Isopropamide, Pipenzolate methylbromide

4. Miscellaneous group of drugs possessing anti-muscarinic effects
➤ **Antihistamines:** Diphenhydramine, Promethazine, Orphenadrine
➤ **Phenothiazine group of antipsychotics:** Chlorpromazine, Thioridazine
➤ **Butyrophenone group of anipsychotics:** Haloperidol
➤ **Tricyclic antidepressants:** Amitriptyline, Imipramine

Pharmacokinetics

- Absorption: these drugs are well absorbed from the gut & across the conjuctival membrane

- Distribution: except quaternary compounds, rest of the drugs gets widely distributed in all body compartments. Scopolamine is rapidly & fully distributed in CNS & has greater effects

- Metabolism: 50% atropine & 80% of scopolamine is metabolized by liver as conjugates

- Excretion: 50% of atropine is excreted unchanged through urine, $t_{1/2}$ is 3 hrs

Pharmacological Actions

- Sensitiveness of different smooth muscles & glands toward atropine action
Sweat, Bronchial & Salivary glands >> Heart & Eye >> Bladder & GIT >> Gastric glands

Central Nervous System

- Scopolamine has the greater permeability through BBB
- Atropine has effect on CNS only in higher doses & toxic doses
- In higher doses → stimulates higher cerebral centres
- In toxic doses → central excitation & leading to restlessness, irritability, disorientation, hallucinations & delirium
- With still large doses, the stimulation followed by depression leading to circulatory collapse, paralysis, coma & respiratory failure leading to death
- Scopolamine in therapeutic doses produces drowsiness, amnesia, fatigue, dreamless sleep (with reduction in REM sleep) & depression of vomiting centre by suppressing vestibular excitation
- In toxic doses, it causes agitation, excitement & hallucinations
- Still higher dose → stimulation is followed by depression leading to coma & respiratory failure

Eye

- They block M_3 receptors in papillary constrictor muscle → produces mydriasis
- The normal papillary responses being blocked, the eyes become unresponsiveness to light (loss of light reflex)
- They also block M_3 receptors at the ciliary muscle of lens → the suspensory ligaments get tightened resulting in flattening of the lens (i.e. becomes less convex) → eye set for distant vision → this effect is termed as paralysis of accommodation or "Cycloplegia"
- Both mydriasis & cycloplegia → precipitates rise in IOP in elderly persons or in individuals with shallow anterior chambers or with narrow angle glaucoma → this effect is due to falling of iris back over the canal of Schlemm which obstructs the drainage of aqueous humor
- They also blocks M_3 receptors at lacrimal glands → decrease in lacrimation → dry or sandy eyes

Cardio Vascular System

- In clinical doses, Atropine causes a transient bradycardia initially → due to M_1 antagonism
- Further dose leads Tachycardia → due to M_2 antagonism on the SA node
- Higher doses of atropine → dilates cutaneous blood vessels especially in face

Respiratory System

- Atropine decreases the secretions of nose, mouth, pharynx & bronchi → dry the mucus membrane of respiratory tract
- Also reduce the laryngospasm during General Anesthesia

- Drying of mucus secretions & suppression of mucociliary clearance → leads to formation of mucus plugs in patients with airway diseases → obstruct the air flow → predispose the patient to infection
- Ipatropium antagonize bronchoconstriction induced by histamine, bradykinin & $PGF_2\alpha$ → they block the indirect effects of inflammatory mediators that are released during the attacks of asthma
- Besides, these drugs have lesser drying effects on sputum (no risk of forming mucus plugs) & don't inhibit mucociliary movements

Gastro Intestinal Tract

- These drugs reduce the basal secretions (fasting phase) than intestinal phase secretions (secretions stimulated due to food, nicotine or alcohol) [M_1 & M_3 antagonism]
- They reduce the tone & motility of the gut from stomach to colon → prolongation of gastric emptying time, closure of sphincters, decrease in tone, amplitude & frequency of peristaltic movements [M_3 antagonism]
- They also posses spasmolytic activity → hey relax the gut in absence as well as in the presence of cholinoceptor stimulants
- These drugs also relax the bile duct & gall bladder

Genitourinary Tract

- They relax the smooth muscles of ureters & urinary bladed wall → voiding is slowed (urinary retention)
Sweat glands

- They cause decrease in sweating → skin becomes dry and hot (rise in body temperature occurs only in higher doses) → Atropine fever

Therapeutic Uses

Motion Sickness

- Scopolamine (0.6mg – 1.0mg s.c.) is effective in motion sickness caused during landing & takeoff by an aero plane or while traveling on a ship or traveling high altitude
- **Basis:** during landing & takeoff by an aero plane or while traveling on a ship or traveling high altitude → due to disparity between sensory inputs received from non-vestibular proprioceptors (muscle spindle, Golgi tendon organs & deep connective tissue receptors) & vision as well as sensory outputs from vestibular apparatus to cerebellum. → results in dizziness, loss of balance, nausea and vomiting
- Diphenhydramine, Cyclizine or Meclizine → prevention of Motion sickness & for treatment of vertigo due to labyrinth dysfunction
- Cinnarizine → antivertigo drug → also used for the prevention of Motion sickness → its antihistaminic, anticholinergic, antiserotonin & Ca^{+2} channel blocking → by inhibiting the influx of Ca^{+2} from endolymph into the vestibular apparatus → blocks labyrinthine reflexes

Parkinson's disease

- Tremors & rigidity associated with Parkinson's disease seem o result from the relative dominance of cholinergic activity in the basal ganglia
- The combination of Antimuscarinic drug with a dopaminergic drug provides more effective
- Muscarinic receptors antagonists are used to treat the extrapyramidal side effects
- The centrally acting anticholinergic drugs are
  - Benzotropine (1-5 mg/day)
  - Benzhexol (2-10mg/day)
  - Procyclidine (5-15mf/day)

Diagnosis of Alzheimer's Disease

- Tropicamide is used to diagnosis the AD (it is instilled in the eyes of patient with suspected AD has been found to exhibit unexpectedly marked dilatation of the pupil → due to changes in receptor sensitivity associated with the disease)

Lie detector

- Hyoscine with morphine produces twilight sleep with amnesia during labour
- It was famous as lie detector agent during World War – II as it produces sedation with amnesia

As Mydriatic

- During the Ophthalmoscopic examination of the retina

To prevent the adhesions in inflammatory conditions

- They are used along with miotics to prevent adhesions between the iris & the anterior surface of the lens as in iridocyclitis, iritis or uveitis

Bronchial Asthma & COPD

- Chronic Obstructive Pulmonary Disease (COPD) → where cholinergic tone is more important contributory factor
- Ipatropium reduces bronchial secretions, least drying effect n sputum, do not interfere with mucociliary clearance & cause bronchodilatation → so they are preferred in COPD
- Ipatropium can be given with β2 agonist it the treatment of Asthma due to longer bronchodilator activity

Preanesthetic Medication

- To reduce bronchial secretions (which is contributory factor in producing laryngospasm) & to prevent excessive vagal effect on heart
- But now a days no body using these drugs
- Scopolamine has an advantage because of its CNS depressants effects (amnesia & tranquillisation)
- Glycopyrolate causes less tachycardia & reduces bronchial , salivary secretions → so it is preferred in preanesthetic medication

Peptic Ulcer

- Proprantheline & Glycopyrolate were more preferred because they don't cross the BBB & less side effects
- These drugs reduce the basal secretions (fasting phase) [M_1 & M_3 antagonism]
- These drugs are not useful in the management of gastric ulcers as they increase the gastric emptying time → prolong the exposure of the ulcer bed to gastric acid
- Pirenzepine & Telenzepine are selective M_1 antagonists
- They are used along with other drugs in the management of peptic ulcer

Antispasmodics

- Methylatropine, Hyoscine methylbromide, Dicyclomine, Propantheline are used in the conditions of hypermotiliy of the gut as in intestinal colic, traveler's diarrhea, irritable bowel syndrome, mild dysentery
- These drugs are also used for the treatment of Biliary colic

To reduce excessive salivation

- These are used to reduce the excessive salivation associated with heavy metal poisoning or parkinsonism

Therapeutic uses related with Genitourinary tract

- Dicyclomine & Oxybutinin are used in the treatment of renal colic & to relieve urethral smooth muscle spasm
- Oxybutinin improves the bladder capacity
- Flavoxate, a directly acting smooth muscle relaxant → which anticholinergic drug used in urinary incontinence & for suprapubic pain in cystitis & urethritis

Miscellaneous Uses

- Treatment of Mushroom poisoning
- Treatment of Muscarinic side effects of neostigmine
- Treatment of Organophosphorus poisoning along with AChE reactivating drugs

Adverse Effects & Toxicity

Most common side effects are

o Dryness of mouth
o Blurred vision & photophobia
o Constipation
o Urinary retention
o Decreased sweating
o Precipitation of glaucoma
▪ Toxicity occurs due to overdosing (above 80 mg orally)

Symptoms of poisoning (Atropine Poisoning)

o Dry skin (as dry as bone)
o Hyperpyrexia
o Flushing of face (as red as beet)
o Mydriasis
o Photophobia (as blind as bat)
o Dry mouth
o Slurred speech
o Difficulty in micturation
o Confusion, delirium
o Hallucinations (as mad as hen)
o Tachycardia

Treatment (Atropine Poisoning)

o The right antidote is Physostigmine (1 – 4 mg slowly i.v every 2 hourly) till satisfactory Muscarinic blockade produced by atropine is countered or overcome
o Basis: Physostigmine is a reversible anticholesterase & chemically a tertiary amine → crosses BBB & indirectly increase the concentration of ACh → it can competitively overcome the antagonistic effect of atropine at Muscarinic & nicotinic receptors all over the body including in CNS
o Other measures
▪ Control of hyperpyrexia by cold sponging, artificial respiration, oxygen by facemask
▪ Removal of unabsorbed drug by gastric lavage & putting universal antidote in stomach
▪ To control excitement & convulsion, give diazepam 2 mg i.m.

Autonomic Nervous System (Adrenergic)

- Noradrenaline is the major neurotransmitter in the peripheral sympathetic nervous system
- Adrenaline is the primary hormone secreted by adrenal medulla
- Dopamine is the metabolic precursor of NAdr & Adr & third neurotransmitter in sympathetic system
- The neurotransmitter at the sympathetic ganglia is Acetylcholine
- All postganglionic sympathetic fibers release Noradrenaline at the Neuroeffector junction

Adrenergic receptors

- Two types of adrenergic receptors – α and β

α – Adrenergic receptors

α_1 adrenergic receptors

- They are present post-synaptically
- When stimulated by agonist these produce excitatory effects in vascular smooth muscle, salivary glands, bronchi, uterus, radial muscle of iris, urinary bladder neck, liver cells
- When stimulated by agonist these produce inhibitory effects in Intestines (relaxation)
- α_{1A} *Receptors:* mainly located in heart, liver, cerebellum, urinary bladder neck & vas deferens
- α_{1B} *Receptors:* mainly located in kidneys & splenic capsule
- α_{1D} *Receptors:* mainly located on hippocampus & prostate

Basis

- The excitatory responses of α_1 receptor activation are mediated through inositol phosphate pathway
- The inhibitory effects of α_1 receptor activation in the gut results when G-protein coupled receptors acts through an increase in the ionic permeability of K^+ ions leading to hyper-polarization

α_2 adrenergic receptors

- They are present pre-synaptically on the sympathetic postganglionic neuron
- Stimulation of these receptors by NAdr decreases further release of NAdr
- Blocked of these receptors produces more NAdr
- Presynaptic α_2 receptors are also located on the cholinergic nerve terminals of the gut → when these receptors are activated, the release of ACh is decreased from the cholinergic neuron leading to relaxation
- α_2 receptors post-synaptically located in brain, in platelets (causing aggregation) & on β-pancreatic cells (inhibiting insulin release)
- α_{2A} receptors are located in platelets, α_{2B} receptors are located in liver & kidney, α_{2C} receptors are found post-synaptically in the cerebral cortex
- In brain, the α_2 receptors whether present post-synaptically or pre-synaptically decrease the sympathetic outflow upon their activation

Basis

- α_2 receptor agonists inhibit adenylate cyclase through an inhibitory G-protein (G_i) & reduce the intracellular cAMP concentration → results in a decreased activation of cAMP dependent protein kinase → diminished response

β – Adrenergic receptors

β_1 – adrenoceptors

56

- These are mainly found in heart → they produce positive inotropic & positive chrontropic effects (i.e. excitatory effects)
- They are also found in kidney → promotes rennin release

β₂ – adrenoceptors

- They are found in bronchi, coronary arteries, uterus & smooth muscles (including gut)
- Their activation causes smooth muscle relaxation (inhibitory effects)

β₃ – adrenoceptors

- These are mainly found on adipocytes → activation of these receptors promotes lipolysis (rise in free fatty acids in the blood) & thermogenesis

Presynaptic β receptors

- These are identified on adrenergic nerve terminals
- Activation of these receptors → facilitates the neurotransmitter release (opposite to that usually observed after pre-synaptic α_2 adrenergic receptor activation

Dopamine Receptors

- Dopamine is a endogenous catecholamine neurotransmitter
- Its actions are mediated through specific dopamine receptors D1, D2, D3, D4 & D5
- At higher doses it activates β_1 receptors and still higher activates α_1 receptors
- D1 appears to be more important in the periphery (renal, mesenteric & coronary vascular beds)
- D2 subtype has effect more in CNS

Basis

- D1 & D5 → stimulation of adenylate cyclase → increase in cAMP → causes vasodilatation in the renal & mesenteric arteries → increased glomerular filtration, renal blood flow & excretion of sodium
- D2, D3 & D4 → inhibits adenylate cyclase → decrease in cAMP → opens K+ channels & blocks Ca+2 channels
- Dopamine receptors causes peripheral vasodilatation in the renal system and also increases myocardial contractility by direct stimulation of β_1 receptors → increases cardiac output but total peripheral resistance & mean BP is unchanged due to simultaneous renal & splanchnic vasodilatation through D1 receptor

Biosynthesis

- L-Phenylalanine → L-Tyrosine → transported into neuronal cytoplasm → converted to L-Dopa → Dopamine → Noradrenaline in storage vesicle → adrenaline into cytoplasm

Pharmacological actions

Eye

- The radial papillary dilator muscle of iris contains β_1 receptors → activation causes Mydriasis
- The ciliary muscle possesses β_2 receptors → activation of these receptors causes very insignificant effect due to dominance of parasympathetic system
- α_1 receptors agonists increases the outflow of aqueous humor → decrease in IOP
- Smooth muscles of eyelid α_1 receptors → stimulation → constriction of eyelid
- Active mydriasis (more light) with no cycloplegia (no blurred vision) is the major advantage over atropine mydriasis

Salivary glands

- Sympathetic activation produces thick salivary secretions → due to reduced blood flow caused by vasoconstriction

Lungs

- Bronchial smooth muscles possesses β_2 receptors → activation causes relaxation (bronchodilation)
- The blood vessels of the mucous membrane of upper respiratory tract (nasal & bronchial mucosa) get constricted by Adrenaline → sympathetic stimulation causes decongestion (α_1 effect)
- Both, bronchodilatation (β_2 effect) & mucosal vasoconstriction (α_1 effect) increases the diameter of lumen & decrease bronchial resistance
- Adrenaline also inhibits the release of allergic mediators from mast cells (β_2 effect) → further reduces congestion

Heart

- Direct effects on heart are mainly by β_1 receptors
- Activation of β_1 receptors at SA node → increases the heart rate
- Activation of β_1 receptors at AV node → enhances conduction & decreases effective refractory period
- Activation of β_1 receptors at Purkinje fibers → increases automaticity
- Activation of β_1 receptors at Ventricular myocardium → increases the force of contraction

Blood Vessels

- Vascular smooth muscle tone is governed by → α_1 & β_2 receptors
- α_1 receptors activation → increase arterial resistance
- β_2 receptors activation → causes smooth muscle relaxation
- The skin, mucosal, splanchic & renal blood vessels → more α_1 receptors effect → constriction during sympathetic stimulation
- But in renal D1 receptors promotes vasodilatation of renal arteries → renal vascular resistance doesn't increases due to α_1 effect
- Coronaries have predominantly β_2, α_2 & D_1 → stimulation → exhibit vasodilatation & also increases myocardial contractility & increases its oxygen demand producing local hypoxia
- Local hypoxia further dilates the coronaries
- α_2 receptors are found in endothelial cells of coronary arteries → stimulation → causes release of NO → causes vasodilatation
- Post-synaptic β_2 receptors are located on skeletal muscle blood vessels → stimulation → causes vasodilatation → reduce in peripheral vascular resistance
- Pulmonary blood vessels have both α_1 & β_2 receptors → stimulation → causes vasoconstriction due to the domination of α_1 receptors

Gastro Intestinal Tract

- Smooth muscles of stomach & intestine have α_2 & β_2 receptors
- α_2 receptors are located presynaptically → stimulation causes → reduces release of ACh → tone & motility reduces
- α_2 receptors → stimulation → decreases salt & water influx into the lumen of intestine
- α_1 receptors → stimulation causes → contraction of sphincters
- β_2 receptors are located directly on smooth muscle → stimulation → causes relaxation via hyperpolarisation

Genitourinary Tract

- Sympathetic stimulation → faster urinary retention → relaxing the detrusor muscle (β_2 effect) & contracting the trigone & sphincter of the urinary bladder (α_1 effect)
- The response of uterus varies with the state of gestation
- Uterine muscle of non-pregnant uterus ($\alpha_1 > \beta_2$) is contracted while of pregnant uterus ($\beta_2 > \alpha_1$) gets relaxed
- Ejaculation is facilitated by activation of α_1 receptors in vas deferens, seminal vesicles & prostate

Skin

- Sweating from palms, sole & forehead is increased during psychologic stress due to sympathetic stimulation
- Pilomotor muscles are contracted (α_1 effect) leading to piloerection (raising of hair)

Metabolic Effects

- The major metabolic effect are, increased blood concentration of glucose, lactic acid & free fatty acids

In adipocytes

- Stimulation β_3 receptors → activates lipase enzyme → catalyses the breakdown of triglycerides to free fatty acids in the fat cells → an increase in oxygen consumption

In liver

- β_2 stimulation catalyses phosphorylase enzyme → promotes glycogenolysis (synthesis of glucose from glycogen) & gluconeogenesis (synthesis of glucose from lactate & amino acids)
- The release of glucose from liver is accompanied by an efflux of K+ ions → there is hyperglycemia with hyperkalemia → K+ released from the liver is taken up by skeletal muscle → followed by hypokalemia

In Skeletal Muscle & Heart

- The glycogenolysis & glycolysis (breakdown of glucose) → produces lactic acid (β_2 effect) → hyperglycaemia

In Pancreas

- α effect dominates over β effect
- predominant action of catecholamines is on α_2 receptors of the pancreatic B-islet cells → diminution in insulin secretion
- Increase in glucagon secretion from pancreatic-A-islet cells through activation of β_2 receptors → hyperglycaemia

Skeletal Muscle

- Stimulation of α-receptor at cholinergic neuron of somatic motor system, increases Ach release ← due to the influx of Ca^{+2}
- β_2 receptor activation improves the blood flow to skeletal muscle (partly counteracted by vasoconstriction of α_1 receptors) → net result is "antifatigue effect" during stressful situations

Kidney

- Increase in rennin release from juxtaglomerular apparatus ← β_1 effect
- Increase in the motility & tone of ureter ← α_1 effect

Other effects

1. Inhibition of histamine release through mast cell membrane stabilization
2. Inhibition of lymphocytic activity
3. More uptake of K^+ from ECF to ICF which results in the fall of plasma K^+ levels (β_2 effect) → decrease in K^+ levels prevents occurrence of arrhythmias due to stress

Type	Tissue	Actions
α_1	Most vascular smooth muscle (innervated)	Contraction
	Pupillary dilator muscle	Contraction (dilates pupil)
	Pilomotor smooth muscle	Erects hair
	Prostate	Contraction
	Heart	Increases force of contraction
α_2	Postsynaptic CNS adrenoceptors	Probably multiple
	Platelets	Aggregation
	Adrenergic and cholinergic nerve terminals	Inhibition of transmitter release
	Some vascular smooth muscle	Contraction
α_2	Fat cells	Inhibition of lipolysis
β_1	Heart	Increases force and rate of contraction
β_2	Respiratory, uterine, and vascular smooth muscle	Promotes smooth muscle relaxation
	Skeletal muscle	Promotes potassium uptake
	Human liver	Activates glycogenolysis
β_3	Fat cells	Activates lipolysis
D_1	Smooth muscle	Dilates renal blood vessels
D_2	Nerve endings	Modulates transmitter release

Therapeutic Classification of Adrenergic Drugs

1. **Pressor agents:** Noradrenaline, Ephedrine, Dopamine, Phenylephrine, Methoxamine, Mephentermine
2. **Cardiac Stimulants:** Adrnaline, Dobutamine, Isoprenaline
3. **Bronchodilators:** Isoprenaline, Salbutamol, Terbutaline, Salmeterol, Formoterol, Bambuterol
4. **Nasal decongestants:** Phenylephrine, Xylometazoline, Oxymetazoline, Naphazoline, Pseudoephedrine, Phenyl propanolamine
5. **CNS stimulants:** Amphetamine, Dexamphetamine, Methamphetamine
6. **Anorectics:** Fenfluramine, Sibutramine, Dexfenfluramine
7. **Uterine relaxant & Vasodilators:** Ritodrine, Isoxsuprine, Salbutamol, Terbutaline

Adrenaline (Epinephrine)
- Primarily secreted by adrenal medulla (80-90%) along with Noradrenaline (10-20%)
- It has equal effects on α and β receptors, $\alpha1=\alpha2$ $\beta1=\beta2$
- At low doses β effects predominate over α effects, At high doses α effects are strong

Therapeutic Uses

Allergy*
- It is the drug of choice in acute hypersensitivity reactions (type 1), like acute ana[hylactic shock → it relieves bronchospasm, angioneurotic oedema of larynx, prevents release of histamine from mast cells & maintains BP

Bronchial Asthma
- It causes bronchodilatation & decongestion of bronchial mucosa

To prolong the duration of action of Local Anesthetic
- Its vasoconstriction effects anatagonises the vasodilating effects of local anesthetic → retards their systemic absorption from the site of administration
- Not only the duration of LA action is prolonged and their systemic toxicity is also decreases

As local Haemostatic
- To control bleeding in Epitaxis & in ENT surgery (used as a spray to have clear vision)

Cardiac Resuscitation
- Intercardiac injection of Adrenaline → used to reverse sudden cardiac arrest like with drowning & electrocution

Adverse Effects
- Increase in BP may lead to cerebral haemorrhage, ventricular tachycardia/fibrillation
- Increase in cardiac work & contractility lead to coronary insufficiency – may precipitate angina, palpitation, arrhythmias
- It induces pulmonary edema
- Transient restlessness, palpitation, anxiety, tremor, pallor
- Contraindicated in Hyperthyroidism → due to up-regulation of α receptors on the vasculature & β receptors in the heart, hyperthyroid individual may become hyper responsive to Adrenaline → dose must be reduced

Dale's Vasomotor Reversal Phenomenon
- Adrenaline produces a biphasic response i.e. rise in BP due to $\alpha1$ & $\beta1$ action (rise in systolic more than diastolic)

Noradrenaline (Norepinephrine)
- Receptor affinity $\alpha1=\alpha2 > \beta1 >> \beta2$
- It raises both systolic & diastolic BP → due to $\beta1$ stimulation (cardiac stimulation) & rise in total peripheral vascular resistance (stimulation of $\alpha1$)
- Reflex bradycardia due to compensatory increase in vagal discharge
- If atropine (which blocks the transmission of vagal effects) prior to Noradrenaline → Hypertension followed with tachycardia
- The coronary blood flow increases (due to rise in mean arterial pressure & also due to production of adenosine – a vasodilator)

Therapeutic uses
- It can be used in cardiogenic shock with precautions (dopamine is preferred)

Adverse Effects
- Sudden fall in BP, necrosis at the site of infusion

Dopamine
- It is a immediate precursor of Noradrenaline & Adrenaline
- It doesn't cross BBB → no central effects

Therapeutic Uses

- Used in Cardiogenic shock resulting from MI, Trauma, Surgery
- Used in the treatment of CHF, renal & liver failure

Rational for the use of Dopamine in Cardiogenic Shock**

- D_1 & D_5 → stimulation of adenylate cyclase → increase in cAMP → causes vasodilatation in the renal & mesenteric arteries → increased glomerular filtration, renal blood flow & excretion of sodium
- D_2, D_3 & D_4 → inhibits adenylate cyclase → decrease in cAMP → opens K^+ channels & blocks Ca^{+2} channels
- Dopamine receptors causes peripheral vasodilatation in the renal system and also increases myocardial contractility by direct stimulation of β_1 receptors → increases cardiac output but total peripheral resistance & mean BP is unchanged due to simultaneous renal & splanchnic vasodilatation through D_1 receptor

Adverse effects
- Nausea, vomiting, tachycardia, ectopic beats, hypertension, cardiac arrhythmias

Dipivefrine*
- **Dipivefrine** is a prodrug of adrenaline.
- It is used to treat glaucoma
- It penetrates upon cornea & it is hydrolysed by esterases present there into Adrenaline
- Adrenaline then acts on receptors in the eye to reduce the amount of *aqueous humour* (watery fluid) that fills the back of the eye, therefore reducing pressure.
- It is available as ophthalmic solution.
- It causes vasoconstriction, decreased aqueous humor production, and decreases intraoccular pressure.
- It is preferred over Adrenaline because of longer & more consistent action and also better ocular tolerance
- Side effects are very rare – headache, burning & stinging on instillation

Dobutamine
- It is a racemic mixture → l-form is a potent α1 agonist , d-form is a potent α1 antagonist & powerful β1 agonist
- Net action is → prominent β1 agonist action → more selective inotropic effect than chronotropic without significant change in peripheral vascular resistance & BP
- It is used in patients of Heart Failure associated with myocardial infarction, Cardiac surgery & for short term management of acute CHF ← Dobutamine increases Cardiac output & stroke volume without affecting HR, peripheral resistance or BP
- Half life is 2 mins → steady state is achieved within 10 min of its IV infusion
- A/Es – Sharp rise in BP, HR, Mycardial oxygen demand → angina may precipitate

Psuedoephedrine
- Most commonly used in common cold for symptomatic relief in nasal congestion
- It's a stereoisomer of ephedrine
- Not a specific α1 agonist drug → causes vasoconctriction in mucosae & skin
- It is used as decongestant of upper respiratory tract, nose & Eustachian tubes
- Its mostly combined with antihistamines, mucolytics, antitussives & analgesics
- Dose: 30 – 60 mg TDS

Clonidine

- α_2 agonist drug which is used in the treatment of Hypertension
- After IV injection (higher plasma concentrations) → produces a transient rise in BP , Oral administration (low plasma concentrations) → fall in BP
- It has lesser intrinsic activity on $\alpha2$ receptors present postsynaptically on vascular smooth muscle, at lower plasma concentrations
- ***Basis of antihypertensive action***
- Mainly stimulates $\alpha2$ receptors present at vasomotor center → central sympathetic flow is reduced → fall in BP & HR
- It binds and stimulates the imidazoline receptors in brain (I_1) → modulate central $\alpha2$ activity → reduces central sympathetic outflow
- Activates presynaptic $\alpha2$ receptors present on sympathetic postganglionic neurons → suppress the release of Noradrenaline from peripheral nerve endings
- Decreased central sympathetic flow → reduces rennin release & decreases renal vascular resistance
- Its is especially used in the treatment of Hypertension with poor renal functions
- After Oral administration, bioavailabilty is 100%

Therapeutic uses

- **Moderate Hypertension:** orally 100µg BD along with a diuretic (to counteract the sodium & water retention which leads to the development of tolerance)
- **To control diarrhea in diabetic patients with autonomic neuropathy:** stimulation of $\alpha2$ receptors in GIT increases Na+ & water retention & inhibits HCO3- secretion → reduces intestinal motility by inhibiting the release of Ach
- **In prophylaxis Migraine:** it reduces cerebral blood flow (25-75µg)
- **Management of Nicotine, Alcohol & Opiate withdrawal**
- **Preanesthetic medication:** its used preoperatively for its sedative, anxiolytic & analgesic effects → to provide heamodynamic stability & to reduce anaesthtic requirements
- It attenuates perioperative stress → reduce perioperative oxygen consumption
- **Menopausal hot flushes**
- **For diagnosis of Phaeochromocytoma**
- **For relief of Pain:** epidurally along with opioids in postoperative, neuropathic & labour pain

Adverse effects: Rebound hypertension, Dry mouth, Seadation, Nasal Stuffiness, Constipation, Impotence, Contact dermatitis

Sibutramine

- Newer antiobesity drug → inhibits the reuptake of both NorAdrenaline & Serotonin
- No useful antidepressant property
- It affects food intake by enhancing serotonergic transmission in the hypothalamus & has tranquillising rather than stimulant property
- It reduces food seeking behavior & decreases quantity of food consumed at any meal
- It stimulates thermogenesis by indirectly activating $\beta3$ receptors in adipose tissue
- A/Es – dry mouth, constipation, anxiety, insomnia, mood swings, chest pain, mild increase in BP & HR
- Dose is 10 mg OD

Alpha Blockers Classification

1) Non-Selective

*   _Reversible:_ Phentolamine, Tollazoline
*   _Irreversible:_ Phenoxybenzamine
2) Selective

*   _α_1 blockers:_ Prazosin, Terazosin, Doxazosin, Alfuzosin, Tamsulosin, Bunazosin
*   _α_2 blockers:_ Yohimbine, Idazoxan
3) Misc: Ergot Alkaloids

Phentolamine

CVS Effects

*   Blockade of vasoconstrictor α_1 receptor in the periphery → leads to vasodilatation → decreases in peripheral vascular resistance → hypotension → stimulates baroreceptor reflex causing → sympathetic discharge → stimulates β_1- receptors on the heart → tachycardia
*   Also block presynaptic α_2- receptor → promotes neuronal release of NE → to produce more tachycardia & palpitation
*   Postural hypotension → due to pooling of blood → cerebral hypoxia, vertigo, fainting
Other P'cological Actions

*   **Nasal Stuffiness:** due to vasodilatation & congestion of nasal mucosa
*   **Miosis:** due to loss of radial muscles of iris & unopposed contraction of circular muscle
*   **Improved urine flow rates:** due to smooth muscle relaxation of bladder neck & prostate
*   **Failure of ejaculation & impotence:** due to inhibition of contractions of vas deferens & ejaculatory ducts
*   **Nausea, vomiting & diarrhoea:** due to partial inhibition of relaxant sympathetic influences on GIT & increase in gastric secretions due to agonist action on H_2 receptors of histamine
Therapeutic uses

*   **For the diagnosis & management of Phaeochromocytoma** [a tumor that is derived from chromaffin cells and is usually associated with paroxysmal or sustained hypertension]
*   **For pheripheral Vascular disorders:** to counteract the α-receptor mediated vasospastic component in the extremities → in Raynaud's disease & Frost bite
*   **To prevent dermal necrosis:** given S.C. to prevent dermal necrosis after incidental extravasation of NE from IV infusion
*   **To treat hypertensive crisis:** follow abrupt withdrawal of clonidine or those resulting from interaction of ingesting tyramine containing food with MAO inhibitors
*   **To treat Male sexual dysfunction:** direct intra cavernous injections of phentolamine (along with papaverine) are given to treat erectile dysfunction → intrapenile injections lead to fibrotic lesions
Phenoxybenzamine

*   It binds covalently to α_1 & α_2 receptors → causing irreversible blockade → longer duration (20-48 hrs)
*   It inhibits the reuptake of released NE by adrenergic nerve terminals → more availability of NE
*   it also blocks receptors for Ach, Histamine & 5-HT

- It crosses BBB
- It causes vasodilatation, progressive decrease in peripheral resistance & Tachycardia
- Tachycardia → due to blocking of reuptake of NE
- Drug exhibits Dale's Vasomotor reversal phenomenon
- No antagonist for these actions → irreversible → specific drug for Pharochromocytoma
- It causes marked postural hypotension with few first initial doses (first dose phenomenon) → because of impairment of the compensatory vasoconstricting response in the periphery
- **A/E-** inhibition of ejaculation, salt & water retention, cardiac arrhythmias, sedation, fatigue, nausea

Therapeutic uses

- **Treatment of Phaeochromocytoma:** it controls episodes of severe hypertension during surgical manipulation of phaeochromocytoma
- **Treatment of BHP:** blockade of α-receptors in prostatic smooth muscle & bladder base decreases obstructive urine flow [selective α1 blockers like terazosin & tamsulosin are preferred]
- **To control manifestations of autonomic hyper-reflexia in patients with spinal cord transection**
- **To treat Haemorrhagic & Endotoxic shock:** it reverts peripheral vasoconstriction & restores perfusion of the tissue & recovery of such patients

Prazosin

- Prototype α_1 receptor antagonists
- It causes peripheral vasodilatation & fall in arterial pressure with lesser tachycardia → because
- It lacks α_2 receptor blocking actions actions → doesn't promote NE release from sympathetic nerve terminals
- It decreases cardiac preload
- Partly because it suppresses sympathetic outflow from CNS
- It is also a potent inhibitor of cyclic phosphodiesterase enzyme → rise in cAMP in vascular smooth muscle → contribute to its vasodilating effect
- Another: rise in concentrations of HDL & decrease in LDL & triglycerides
- Selective α_1 agonist → it relaxes → smooth muscle in the bladder neck, prostate capsule & prostatic urethra → improves urine flow in BHP

Therapeutic uses

- **Treatment of Hypertension:** α1 receptor antagonists improve lipid profile
- **Dose :** 1mg given at bed time to reduce the risk of postural hypotension → then dose is titrated upward depending on BP
- **To treat BHP :** Dose: 1-5 mg twice daily
- **In Raynaud's disease:** it decreases the incidence of digital vasospasm [but Ca+2 channel blockers are more preferred]
- **In patients with Mitral and Aortic Valvular insufficiency:** it decreases cardiac afterload

Adverse Effects

- Marked Postural Hypotension at initial dose
- Impotence, nasal decongestion, GIT upset & Na+ & water retention

Terazosin & Doxazosin

- These are longer duration of action drugs
- Terazosin ($t_{1/2}$) – 12 hrs, Doxazosin ($t_{1/2}$)-20 hrs[
- Terazosin is more effective than Finasteride as its faster action to improve urine flow

Tamsulosin

- It is α_1 receptor antagonist → further **selectivity for α_{1A} subtype(uroselective)** → which is located on bladder neck, urethra [α1B located mainly on blood vessels]
- It is more efficacious for the treatment of BHP
- T1/2 – 8 hrs
- Sustained release capsules used for once daily dosing
- No cardiovascular side effects
- Side effects - Failure of ejaculation, dizziness

BHP

- The urinary obstruction due to increased size of prostate & due to increased tone of bladder neck/ prostate smooth muscle
- **α1 – blockers** → decrease tone of prostatic/bladder neck muscles (Prazosin 1–5 mg twice daily /Terazosin 1–10 mg daily /Tamsulosin 0.4 or 0.8 mg daily) → relaxation of smooth muscle → reduces dynamic obstruction → increasing urinary flow rate → complete emptying of bladder
- Tamsulosin is preffered → α_{1A} subtype(uroselective) → which is located on bladder neck, urethra → no CVS effects
- **5-α reductase inhibitor** → arrest growth/reduce size of prostate (Finasteride) → blocks the conversion of testosterone to dihydrotestosterone → impacts upon the epithelial component of the prostate → resulting in reduction in size of the gland and improvement in symptoms.
- Six months of therapy is required for maximum effects on prostate size (20% reduction) and symptomatic improvement.
- dose is 0.5 mg orally daily.

Beta Antagonists

- **Non-selective (β1+ β2) blockers:** Propranolol*, Timolol, Sotalol, Nadolo, Pindolol**, Oxyprenolol**
- **Selective Beta Blockers**
- **β1-Blockers:** Metoprolol*, Atenolol, Esmolol, Acebutolol**, Celiprolol**, Betaxolol
- **β2-Blockers:** Butoxamine
- **Mixed (α+β) Blockers:** Labetalol*, Carvedilol

Drugs having membrane stabilising activity

**Drugs having membrane stabilising & Intrinsic sympathomimetic action (β₁ agonist)*

Propranolol

- Prototype non-selective β-blocker
- Not used to treat Galucoma
- β-blockers have little effect on normal resting heart → but exhibit profound action when sympathetic drive to the heart is high (physical exercise or stress)

CVS Effects

- Blockade of β_1 receptors → ↓HR, ↓Myocardial contractility, ↓conduction velocity, ↓C.O.
- Because of ↓ in the workload by heart→ Myocaridal O_2 demand is reduced
- Automaticity of the heart is also ↓ → because of inhibition of latent pacemakers → ↓ B.P
- The radial papillary dilator muscle of iris contains β_1 receptors → activation causes Mydriasis
- The ciliary muscle possesses β_2 receptors → activation of these receptors causes very insignificant effect due to dominance of parasympathetic system
- α_1 receptors agonists increases the outflow of aqueous humor → decrease in IOP
- Smooth muscles of eyelid α_1 receptors → stimulation → constriction of eyelid

- Use of Propranolol → gradual fall in BP → cardiac β_1 –blockade

Other Mechanisms

- ↓renin release → ↓production of angiotensin II & Aldosterone from adrenal cortex
- ↓ Sympathetic flow (central)
- Chronic ↓ C.O
- Blockade of the facilitatory effect of presynaptic β_2 receptors on NE release
- CV effects of Propranolol can be overcome by Isoprenaline → competitive blocker
- On long term use → propranolol lipid profile is worsened; Total triglycerides & LDL-↑, HDL- ↓
- Membrane stabilising (LA action)
- Cardiac depressant actions

Respiratory Effects

- Airways resistance is least affected by β-blockers (Sympathetic bronchodilator tone is less)
- But propranolol → severe bronchoconstriction

Metabolic effects

- Inhibits stress induced or Adr induced glycogenolysis

CNS

- Sedation, lethargy, disturbances in sleep
- Suppresses performance anxiety

MISC

- ↑synthesis & release of PGs
- It prevents aggregation & promotes fibrinolysis (use under bleeding from oesophageal varices)

Pharmacokinetics

- Completely absorbed orally → low bioavailability → first pass metabolism
- It is highly lipid soluble & crosses BBB
- $T_{1/2}$ → 4-6 hrs

Therapeutic Uses

- **Essential Hypertension:** used alone or combined with either a diuretic or a α-blocker
- **Ischemic Heart Disease:** ↓Work load on the heart, ↓myocardial O_2 demand, ↑Exercise tolerance
- **MI:** ↓the incidence, recurrence & mortality in case of MI
- Prevention of platelet aggregation
- **CHF:** β-blockers can worsen heart failure
- These drugs retard the progression of CF → prolong life
- **Cardiac Arrhythmias:** Effective in all supraventricular tachycardias associated with catecholamines
- **Obstruction of Ventricular Outflow:** it provides symptomatic relief from congestion & dyspnoea by reducing contractility & slowing ventricular ejection
- **Oesophageal variceal bleeding & hepatic portal hypertension:** ↓C.O. (β1 blockade) & induce splanchic Vasoconstriction (β2 blockade) → ↓bleeding
- Portal pressure ↓by 40%
- **Hyperthyroidism:** ↓unpleasant symptoms of symp overactivity (tachycardia, irregular cardiac rhythm, tremors, sweating, ↑BMR)
- Also inhibit the peripheral conversion of Thyroxine to Triiodothyronine

Adverse Effects & Contraindications

- *Bronchoconstriction:* Patients with Asthma, Bronchitis or Emphysema
- Contraindicated in Bronchial Asthma or Chronic Obstructive Lung Disease
- *Cardiac Failure:* Cardiac patients always need a degree of sympathetic drive to support the heart to maintain proper cardiac output
- β- blockers block this supposrt & may aggravate heart failure
- Contraindicated in Heart Failure
- *Hypoglycaemia:* The sympathetic responses to Hypoglycaemia (release of glucose through glycogenolysis) → is a safety device that warns the patients of Diabetes [Symptoms like Tachycardia, Cold sweating, tremors]
- ↑in Glucagon secretion from the pancreatic A-islet cells through β2 activation → Hyperglycaemia
- β-blockers mask these symptoms & delay recovery from hypoglycaemia → hypoglycaemic coma
- Contraindicated in patients receiving Insulin & Hypoglycaemic drugs
- *Bradycardia:* Lowering of HR to 55 beats/min → leads life threatening Bradyarrythmias & heart block
- *Cold Extremities:* results from a loss of β2 receptor mediated cutaneous vasodilatation in the extremities
- Contraindicated in Raynaud's disease
- *Rebound Hypertension & Anginal attacks after sudden withdrawal:* After chronic therapy → up-regulation in the number of β receptor → they show hyper-responsiveness
- To avoid → should be a gradual tapering of the doses instead of abrupt withdrawal
- *Adverse Serum Profile*
- *CNS Side Effects:* Fatigue (due to ↓C.O. & poor muscle perfusion), sleep disturbances bad dreams, depression
- *Male → sexual dysfunction*

Timolol

- Orally absorbed, moderate 1st pass metabolism, crosses BBB
- Main use: as eye drops to decrease the raised intraocular pressure in cases of wide angle glaucoma
- It is devoid of membrane stabilizing activity

Sotalol

- Low lipid solubility (lesser CNS effects)
- An additional K^+ channel blocking & class-III antiarrhythmic property independent of β -blocking activity

Pindolol & Oxprenolol

- They posses inherent intrinsic sympathomimetic action on $β_1$ & $β_2$ receptors

Though controversial, the benefits of this property are

- Lesser bradycardia & myocardial depression → useful in patients prone to bradycardia or with low cardiac reserve in CHF
- Lesser troublesome in asthmatic patient due to $β_2$ agonist action
- Rebound hypertenion after withdrawal is less likely as β -agonist action → prevents the supersentivity of up-regulated β-receptors
- Lipid profile is less worsened, compared to Propranolol

Disadvantage of Sympathomimetic activity

- These drugs can't be used in migraine prophylaxis as intrinsic $β_2$-agonist activity dilates cerebral blood vessels
- Less suitable for secondary prophylaxis of MI

Selective β-Blocker (Cardio Selective)

- *Metoprolol, Atenolol, Esmolol, Betaxolol*
- In therapeutic doses block β_1 receptors, in high doses it can block β_2 receptors

The advantages over non-selective blockers

- These are safer in asthmatics, compared to propranolol → they are not absolutely safe
- These are safer in diabetes as they cause less inhibition of glycogenolysis during hypoglycaemia (glucose release is controlled by β_2 receptors)
- These have deleterious effects on lipid profile
- These are safer in patients with peripheral vascular disease (no β_2 blockade)

Disadvantages of Cardioselective blockers

- Rebound hypertension after abrupt withdrawal of the drug
- These are ineffective in controlling essential tremors

Esmolol: Ultra-short acting (t1/2 – 8-10 min)

- Used – to terminate supraventricular rachycardia, episodes of atrial fibrillation, to control HR & BP during surgery , or in critically ill patients in whom the drug can be withdrawn immediately if adverse effects like bradycardia, hypotension or heart failure appear

Labetolol(Mixed)

- Competitive antagonist of both α_1 & β-recpetors
- Its actions are
- Selective blockade of α_1- receptor & Blockade of β_1-receptor → fall in BP
- Blockade of β_2- receptor
- Partial agonist activity (ISA) at β_2 receptor → peripheral vasodilatation & bronchodilatation
- Inhibition of neuronal uptake of NE
- It has direct vasodilating capacity
- HR is unaffected
- Postural hypotension & hepatotoxicity are the main side effects
- Orally effective, undergoes 1st pass metabolism, Bioavailabilty 30% , $t_{1/2}$ – 4-5 hrs
- Used in – Hypertension in elderly patients where increased peripheral resistance is not desired
- Particularly useful in phaeochromocytoma & for controlling rebound hypertension after clonidine withdrawal

1. Therapeutic Classification of Adrenergic agonist drugs. Enumerate the therapeutic uses and adverse effects of Adrenaline (Epenephrine)
2. Give the pharmacological basis (or) rational for the use of
a. Adrenaline in Anaphylactic shock
b. Dopamine in adrenergic shock
c. Pseudoephedrine as nasal decongestant
d. Tamsulosin in BHP
e. Adrenalin along with a Local Anesthetic
f. Adrenergic drugs in Obesity
g. Propranolol in Throtoxicosis
h. Bisoprolol in MI
i. For contraindications of Propranolol in diabetic patient
3. Short notes on
a. Treatment of BHP
b. Prazosin
c. Dipivefrin
4. Differences among
a. Beta Blockers
5. List
a. Adverse effects of Beta blockers
b. Adrenergic receptor drugs

Cardio Vascular System

Anti-anginals, Antihypertensives, CHF, Antiarrhythmics

Anti-Anginal Drugs

[**Angina Pectoris** - Pain occurred due to lack of blood & O_2 to Myocardium]

Pathophysiology of Angina & Types of Angina
- Angina pectoris, the primary symptom of ischemic heart disease, is caused by transient episodes of myocardial ischemia that are due to an imbalance in the myocardial oxygen supply-demand relationship → This imbalance may be caused by
    - An increase in myocardial oxygen demand (which is determined by heart rate, ventricular contractility, and ventricular wall tension) or
    - By a decrease in myocardial oxygen supply (primarily determined by coronary blood flow but occasionally modified by the oxygen-carrying capacity of the blood) or
    - Sometimes by both
- The progressive decrease in vessel radius → characterizes coronary atherosclerosis can impair coronary blood flow → lead to symptoms of angina when myocardial oxygen demand increases, as with exertion (so-called **typical angina pectoris** → occurs during exercise, emotions).
- In some patients, anginal symptoms may occur without any increase in myocardial oxygen demand but rather as a consequence of abrupt reduction in blood flow ← result from coronary thrombosis (**unstable angina** → occurs in exercise & in rest) or vasospasm (**variant or Prinzmetal angina** → occurs in rest).
- Regardless of the precipitating factors, the sensation of angina is similar in most patients.
- Typical angina is experienced → as a heavy, pressing substernal discomfort (rarely called "pain"), often radiating to the left shoulder, flexor aspect of the left arm, jaw, or epigastrium.
- A significant minority of patients notes discomfort in a different location or of a different character.
- Women, the elderly, and diabetics are more likely to have ischemia with atypical symptoms.

Aim/Goal of therapy
- Provide prophylactic or symptomatic treatment
- Decreasing the Oxygen demand by the heart & uniformly distribution of the blood which is available
- Decreasing the load on the heart by decreasing pre-load & after-load
- Reduce mortality apparently by decreasing the incidence of sudden cardiac death associated with myocardial ischemia and infarction
- The treatment of cardiac risk factors can reduce the progression or even lead to the regression of atherosclerosis

Classification of drugs use in the treatment of Angina Pectoris

1. **Nitrates**
    - *Rapid onset, short acting:* Nitroglycerine, Amyl Nitrate
    - *Slow onset, long-acting:* Isosorbide dinitrate(sublingually → its short acting), Isosorbide mononitrate
2. **Calcium Channel Blockers (CCBs) :** Verapamil, Diltiazem, Nifedipine, Nicardipine, Nimodipine, Amlodipine
3. **Beta Blockers :** Propranolol, Atenolol, Metoprolol, Nadolol
4. **Miscellaneous**
    - *Potassium Channel Openers:* Nicorandil
    - *Cytoprotective Drugs:* Trimetazidine, Dilazep
    - *Antiplatelet Drugs:* Aspirin, Ticlopidine, Clopidogrel, Dipyridamole
    - *Drugs of Lesser Importance:* Lidoflazine, Oxyfedrin, Phentoxiphylline

Mechanism of Action of Antianginal Drugs

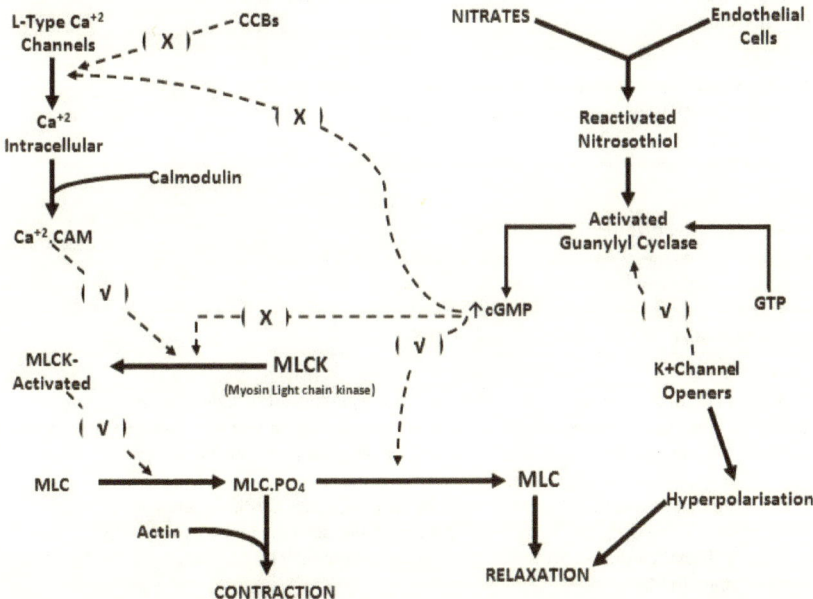

Nitrates

Mechanism of Action

- Nitrates are rapidly denitrated enzymatically in smooth muscle cell to release the reactive free radical *Nitric Oxide (NO)* → activates Cytosolic Guanylyl Cyclase → increased cGMP → dephosphorylation of MLCK through a cGMP dependent protein kinase
- Reduced availability of MLCK-P interferes with activation of Myosin → fails to interact with Actin to cause contraction
- Raised intracellular cGMP also reduces Ca^{+2} entry → helps in relaxation

Pharmacological Actions

Preload Reduction

- Nitrates exerts prominent action on vascular smooth muscle
- Nitrates dilate veins more than arteries → peripheral pooling of blood → decreased venous return i.e. preload on heart is reduced → decreased cardiac work → reduced O_2 consumption.

After load Reduction

- Nitrates also produce arteriolar dilatation → Decreases peripheral resistance → systolic BP falls more
- reduction in cardiac work

Redistribution of coronary flow

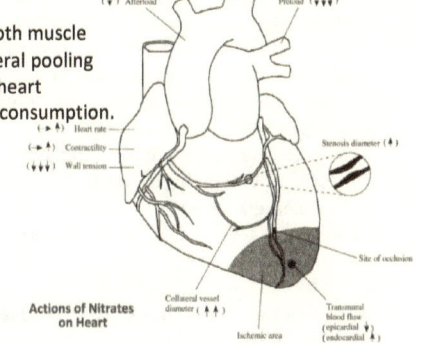

Actions of Nitrates on Heart

73

- Nitrate relax bigger conducting coronary arteries →
redistribution of blood flow to ischaemic areas in angina patients

Pharmacokinetics

- Lipid soluble, well absorbed from buccal mucosa, intestines, skin
- Undergo extensive First Pass Metabolism (except mononitrate)
- Longer $t_{1/2}$

Adverse Effects: these are mostly due to vasodilatation

- Fullness in head, throbbing headache
- Flushing, weakness, sweating, palpitation, dizziness & fainting
- Methemoglobinemia
- Rashes

Uses of Nitrates

1. **Angina Pectoris:** effective in classical & variant angina (unstable angina → antiplatelets)
2. **CHF & acute LVF:** Nitrates afford relief by venous pooling of blood → reduced venous return (preload) → decreased end diastolic volume → improvement in left ventricular function & regression of pulmonary congestion
3. **Myocardial Infarction:** i.v. infusion of GTN to avoid tachycardia → area of necrosis can be reduced by altering O_2 balance in the marginal partially ischaemic zone by reducing cardiac work
4. **Interventional cardiac procedures:** such as percutaneous coronary angioplasty, thrombolytic therapy of acute MI
5. **Biliary colic:** due to disease or morphine induced biliary colic → GTN or dinitrate
6. **Esophageal spasm:** Sublingual GTN relieves pain → Nitrates before a meal facilitate feeding in esophageal achalasia by reducing esophageal tone
7. **Cyanide poisoning:** Nitrates along with Sodium thiosulfate is used in the treatment of cyanide poising
    - Cyanide actively converts into Cn⁻ ion → chelates cytochrome oxidase iron → which causes tissue anorexia, seizures & death
    - Whereas hemoglobin is converted into methemoglobin in presence of nitrates
    - Methemoglobin has more affinity towards Cn⁻ ion than iron and forms a complex with cyanide ion and forms Cyanomethemoglobin (which is very less stable complex) → Cyanomethemoglobin in presence of Sodium thiosulfate is converted into very strong stable complex Sod thiocyanate and methemoglobin → sodium thiocyanate is excreted through urine

β-Blockers

- They don't dilate coronaries or other blood vessels → total coronary flow is reduced due to blockade of dilator β_2 receptors
- They act by reducing heart rate, myocardial contractility (β_1 blockade) → results in decrease in cardiac output & decreased myocardial O_2 requirement, at rest & during exercise
- increases the magnitude of ST segment depression in ECG during exercise
- All β-blockers are equally effective in decreased frequency & severity of attacks & increases exercise tolerance in Classical Angina
- The blood flow to ischaemic subendocardial area is not reduced due to favorable

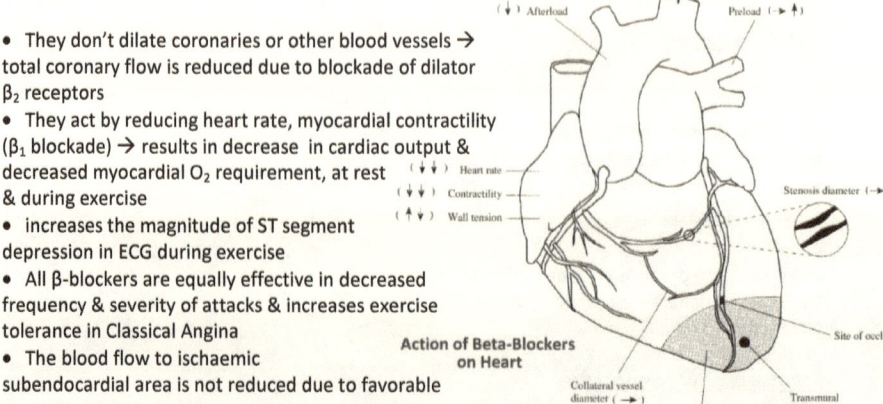

Action of Beta-Blockers on Heart

redistribution & decrease in ventricular wall tension →this is due to increase of diastolic perfusion time
- Beta Blockers inhibit platelet aggregation also
- Higher doses can worsen classical angina by promoting ventricular dilatation & prolonging systole, both of which tend to increase O_2 demand
- They should not be used alone in unstable angina, because of risk of worsening coronary vasospasm
- Abrupt withdrawal → may precipitate an anginal attack and acute MI due to sudden increase in sympathetic tone of the heart
- Cardioselective β_1 blockers like metoprolol or atenolol are more preferred
- Non selective β-blockers like propranolol & nadolol can worse variant angina → because unopposed α-receptor mediated coronary vasoconstriction aggravate coronary spasm
- Higher dose may even worsen classical angina by causing ventricular dilatation & prolong systole → both increases the O_2 demand

Calcium Channel Blockers

- **Phenylalkylamines:** Verapamil
- **Benzothiazepines:** Diltiazem
- **Dihydropyridines**
  - *Short Acting ($t_{1/2}$ 2-6 hrs):* Nifedipine, Nicardipine
  - *Intermediate Acting ($t_{1/2}$ 8-12 hrs):* Nitrendipine
  - *Long Acting ($t_{1/2}$ 30-50 hrs):* Amlodipine

Mechanism of action

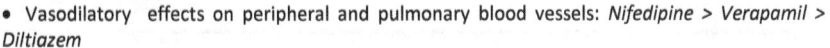

- Verapamil is cardioselective, Nifedipine is vascular smooth muscle selective, Diltiazem exhibits intermediate selective
- Cardiac depressant effects (decreases heart rate, contractility, conduction velocity) :
Verapamil > Diltiazem > Nifedipine
- Coronary artery dilator effects (prevents coronary spasm) :
Nifedipine > Verapamil > Diltiazem

Action of CCBs on Heart

- Vasodilatory effects on peripheral and pulmonary blood vessels: *Nifedipine > Verapamil > Diltiazem*
- All the 3 subclasses of CCBs inhibit L-type Ca^{+2} channel → reduces the frequency of their opening in response to depolarization → decrease in transmembrane Ca^{+2} current in smooth muscle, cardiac muscle, SA node & AV node

Verapamil

- Peripheral vasodilating effects of verapamil are acompanied by cardiac depressant efects (decreases heart rate, conduction velocity & contractility)
- it possesses a nonspecific α-blocking action → peripheral vasodilatation
- Reflex tachycardia occurs due to fall in BP → but it is blunted by the direct negative chronotropic effect
- It is mainly used in mild to moderate hypertension, supraventricular tachyarrhythmias
- the negative inotropic action is useful in the treatment of hypertropic cardiomyopathy
- it is also used to suppress noctural leg cramps
- It is also used to reverse the resistance of cancer cells to some anticancer drugs

Diltiazem

- It is less potent coronary & peripheral vasodilator than Nifedipine or Verapamil

- The suppression of automaticity at SA node & conductivity at AV node are of the same magnitude
- The decrease in the contractility is lesser than Verapamil
- It is used in angina & hypertension

Dihydropyridine CCBs

- *Nifedipine* analogs appear to block vascular smooth muscle Ca^{+2} channels at a concentration below that required for cardiac depressant effects → these are less cardiac depressants
- these are effective as antihypertensive drugs → the fall in BP is accompanied by reflex tachycardia
- The slower & longer acting preparations like amlodipine induce less reflex sympathetic stimulation → lesser tachycardia & lesser tendency to increase cardiac work
- with these CCBs mortality has increased in post MI patients → so widely used (sustained release preparations)
- *Amlodipine* has longer $t_{1/2}$ → given orally once daily
- It has lesser first pass metabolism & bioavailability is consistent
- It has slow but complete oral absorption
- Side effects related to vasodilatation, such as tachycardia, headache & flushing are larely avoided
- It is used for angina and Hypertension

Adverse Effects

- Most common A/Es are due to Vasodilatation – Headache, dizziness, facial flushing, fall in BP
- **Nifedipine** causes reflex tachycardia, ankle oedema, difficulty in voiding urine, hamper diabetes by decreasing insulin release
- **Verapamil** → no tachycardia, but bradycardia is observed, marked negative inotropic effects & depression of AV nodal conductivity
- High doses causes constipation because of relaxant effects on GIT smooth muscles

Beneficial Combinations

➤ *Beta-Blocker + Long acting nitrate combination is rational in Classical angina*
✓ Tachycardia due to nitrtae is blcoked by β-blocker
✓ The tendency of β-blocker to cause ventricular dilatation is counteracted by nitrate
✓ The tendency of β-blocker to reduce total coronary flow is opposed by nitrate
➤ *Nitrates + CCBs*
✓ Nitrates primarily decrease preload
✓ CCBs decreases afterload
✓ Combinely decrease cardiac work which can't achieved by alone drug
➤ *Nitrates with Nifedipine* → advocated for patients of exertional angina with heart failure or sick sinus syndrome or AV nodal conduction defect but excessive tachycardia can be observed
  - *Nitrates primarily decrease preload*
  - *CCBs mainly reduce afterload + increases coronary flow*
  - *beta-blockers decrease cardiac work by direct action on heart*

Potassium Channel Openers

- K^+ ions control the resting membrane potential → K^+ channel opening → efflux of K^+ & Hyperpolarisation → opposes the opening of Ca^{+2} channels (due to no depolarization) → fall of cystolic Ca^{+2} conc resulting in a reduction of cellular contractile activity at the myocardial & vascular level

Voltage Gated K^+ channels

76

- These channels open when the cell is depolarises
- present mainly in vascular & other smooth muscles
- These produce relaxant effects on vascular smooth muscle & other smooth muscles (tracheal, broncheal, intestinal, uterine, urinary bladder & corpus cavernosum

Calcium Activated K⁺ channels
- Increase in intracellular Ca^{+2} con causes opening of these channels → facilitates an influx of K⁺ → repolarisation

ATP sensitive K+ channels
- present in cardiac muscle & β-cells of islets of Langerhans of pancreas
- these are open ATP levels drops
- opening of these channels causes hyperpolarisation & relaxation of cardiac smooth muscle

Nicorandil

- Newer antianginal drug → which activates ATP sensitive K⁺ channels & hyperpolarises vascular smooth muscle
- it contains a nitrate like moiety → it also exerts nitrate like effect→ stimulates cGMP conc in epicardial arteries
- MOA: activation of K⁺ channel (arterial vasodilation) with Nitirc Oxide donor action (venodilator effect)
- Nitrates tolerance doesn't develop here
- It increases coronary blood flow without causing coronary steal phenomenon & without producing significant cardiac effects on contractility, conduction, heart rate & BP
- Adverse Effects: Flushing, palpitation, dizziness, headache, stomatitis, nausea & vomiting
- it is contraindicated in patients with cardiogenic shock, LVF with low filling pressure & hypotension

Myocardial Infarction

- MI results from prolonged myocardial ischaemia
- Myocardial infarction (MI or AMI for acute myocardial infarction), commonly known as a heart attack, occurs when the blood supply to part of the heart is interrupted causing some heart cells to die.
- This is most commonly due to occlusion (blockage) of a coronary artery following the rupture of a vulnerable atherosclerotic plaque, which is an unstable collection of lipids (like cholesterol) and white blood cells (especially macrophages) in the wall of an artery.

Management of Myocardial Infarction
- Thrombolytic therapy reduces mortality & limits infarct size in patients with acute MI
- More benefits if therapy starts within the first 3-4 hrs
- Six fibrinolytic (plasminogen activators) are available to achieve reperfusion of infarcted area
- *Streptokinase, Urokinase, Anistreplase, Alteplase, Reteplase, Tenectplase*
- *Contraindications* of thrombolytic therapy → intracranial hemorrhage, active internal bleeding

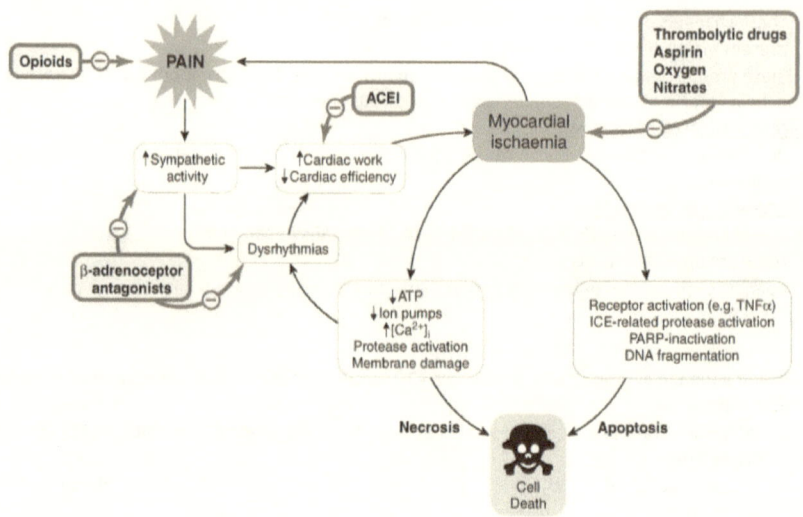

- Streptokinase: not commonly used since it is less effective at opening occluded arteries & less effective at reducing mortality
- it is non-fibrin specific, causes depletion of circulating fibrinogen, induce hypotension → these can be avoided by slow administration
- **Alteplase:** recombinant tissue plasminogen activator, it is fibrin specific
- **Reteplase:** longer duration of action
- **Tenecteplase:** more fibrin sensitivity & more resistance to plasminogen activator inhibitor
- Thrombolytic agent should be administered within 30 mins or during transportation

Post-Thrombolytic Mgt: After completion of the thrombolytic infusion. Aspirin (81-325 mg/day) & heparin should be continued (for 8 days)

- *Low Molecular Weight Heparin in MI:* These can be used to secondary thromboembolic complications, they are given for shorter duration till the patient becomes ambulatory
- Along with Aspirin it reduces the chances of occurring MI
- Chewable Aspirin provides more blood levels
- It found useful in reduction in early death, myocardial reinfarction & stroke, with no excess in major bleeding
- after that sublingual GTN is administered
- *to treat Pain, anxiety, apprehension:* Morphine, Diazepam

    Morphine IV is given in MI and it acts by
    - Due to direct action on the blood vessels → decreases the tone → causes peripheral vasodilatation & also pooling of the blood → reduces preload on heart
    - Again due to vasodilatation → blood shifts from pulmonary to systemic circuit → relieves pulmonary congestion & edema
    - It decreases the pain (due to the action on central μ-receptors which mediate the pain)
    - It decreases the anxiety → by depressing sympathetic flow → calming the patient → reduces cardiac work
    - It is also indicated to relieve pulmonary edema due to infarction of lung & other causes but not due to irritant gases
- Artificial Respiration
- Slow i.v. infusion of Saline / dextran *(to maintenance of Blood Volume)*
- Sod. Bicarbonate by i.v. infusion *(for correction of acidosis produced by lactic acid)*
- Prophylactic administration of β-blockers and continuation of therapy *(to reduce the infarct size, prevents arrhythmias)*

- To increase the cardiac work without increasing the cardiac work
- *Furosemide:* it decreases the cardiac preload
- *Vasodilators:* like GTN, Captopril, Nitroprusside
- *Inotropic agents:* dopamine (*to augment the pumping action*)
- Heparin and followed by oral anticoagulants (*to prevent further formation of thrombus*)
- Fibrinolytic agents (Streptokinase, Urokinase)
- ACE inhibitors (*to prevent the occurrence of CHF*)
- Aspirin on long term basis (*platelet function inhibitors*)
- β-blockers (*to reduce risk of reinfarction, CHF, mortality*)
- Hypolipidemic drugs (*Control of hyperlipidaemia*)

Role of each drug in the treatment of Angina

- **Antianginal agents(Nitrates)** → provide prophylactic or symptomatic treatment
- **Beta blockers** → reduce mortality → by decreasing the incidence of sudden cardiac death associated with myocardial ischemia and infarction → the treatment of cardiac risk factors can reduce the progression or even lead to the regression of atherosclerosis.
- **Aspirin** is used routinely in patients with myocardial ischemia → daily aspirin use reduces the incidence of clinical events
- Other antiplatelet agents such as oral *clopidogrel* and intravenous anti-integrin drugs such as *abciximab, tirofiban,* and *eptifibatide* → reduces morbidity in patients with angina who undergo coronary artery stenting
- Lipid-lowering drugs such as the **statins** reduce mortality in patients with hypercholesterolemia with or without known coronary artery disease
- **Angiotensin-converting enzyme (ACE) inhibitors** → reduce mortality in patients with coronary disease
- Coronary artery bypass surgery and percutaneous coronary interventions such as angioplasty and coronary artery stent deployment can complement pharmacological treatment.
- Percutaneous or surgical revascularization may have a survival advantage over medical treatment alone.
- Intracoronary drug delivery using drug-eluting coronary stents represents an intersection of mechanical and pharmacological approaches in the treatment of coronary artery disease.
- Novel therapies that modify the expression of vascular or myocardial cell genes eventually may become an important part of the therapy of ischemic heart disease.

Calcium Channel Blockers (CCBs)

- The primary use of these drugs → in the treatment of angina, selected cardiac arrhythmias, hypertension
- CCBs → potent vasodilating drugs → lack of fluid accumulating properties like other vasodilators
- 2nd generation analogues → Nimodipipne, Nicardipine, Felodipine, Nisoldipine, Amlodipine → they differ from Nifedipine in their potency, pharmacokinetic characteristics, selectivity of action
- Nimodipine → selectivity for the cerebral vasculature
- Amlodipine → exhibits very slow kinetics of onset and off set of blockade

Calcium Antagonism

- These drugs inhibited Ca+2 component of the ionic currents carried in the cardiac action potential → so these drugs also called 'Slow Channel Blockers"
- Ca+2 → Fundamentally important as a messenger → linking cellular excitation & cellular response ← this is made due to the availability of specific high affinity Ca+2 binding proteins (eg. Calmodulin)
- Ca+2 binding proteins → they serve as intracellular Ca+2 receptor → by the existence of Ca+2 specific influx, efflux, sequestration processes
- Calcium in excess → serves as a mediator of cell destruction & death during myocardial & neuronal ischemia, neuronal degeneration, Cellular toxicity → control of excess Ca+2 mobilisation is an important contributor to cell & tissue protection
- The available Ca+2 channel blocers → exert their effects primarily at voltage gated Ca+2 channel of the plasma membrane
- Types of Ca+2 channels – L, T, N, P/Q, R

Cellular Calcium Regulation

1. Na+, Ca+2 exchange

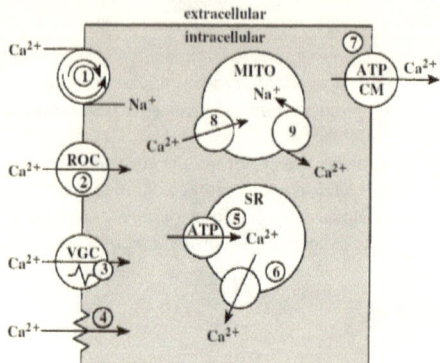

2. Receptor-operated channels.

3. Voltagegated channels.

4. Leak pathways.

5, 6. Entry and efflux in sarcoplasmic reticulum

(ATP - adenosine triphosphate)

7. Plasma membrane pump (CM - cell membrane)

8, 9. Entry and efflux in mitochondria

L-Type Ca+2 channel → 3 receptor sites

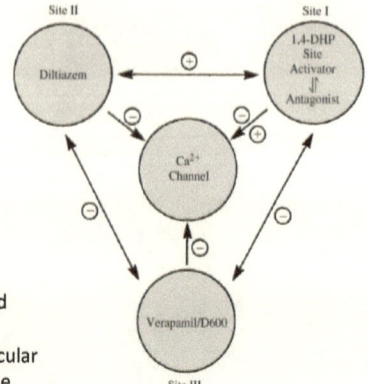

- The activity of Ca+2 channel blocker → increases with increasing frequency of stimulation or intensity & duration of membrane depolarization → this use-dependent activity is consistent with a preferred interaction of the antagonists with the open or inactivated states of Ca+2 channel rather with the resting phase
- Verapamil & Dilatiazem → equipotent in cardiac & vascular Smooth muscle → preferentially act on open channel state

- Nifedipine → more active in vascular smooth muscle → act
Through inactivated state
- Nifedipine & 1,4-DHP analogues → have both agonist & antagonist
properties
- Nifedipine & 1,4-DHP analogues → they are majorly have action on
other than L-type Ca+2 channels → used in variety of CNS disorders,
including neuronal damage & death from ischemic insults
- Verapamil & lesser extent to Diltiazem → posses a number of receptor blocking properties,
together with Na+, K+ channel blocking activities → it may be helpful in their pharmacological
profile
- Nifidipine & other 1,4-DHP analogue → more selective for voltage gated Ca+2 channels →
also affect

Anti-Hypertensive Drugs

Renin-Angiotensin System

- Angiotensin II generated in plasma from plasma α_2 globulin → involved in electrolyte, blood volume, pressure homeostasis

Cardio Vascular System
- Vasoconstriction (direct action) & enhancing Adrenaline/Noradrenalin release from adrenal medulla/adrenergic nerve endings → increases central sympathetic outflow
- Increased Myocardial contraction by promoting Ca^{+2} influx, increased Heart rate by enhancing sympathetic activity, reflex bradycardia, decreases Cardiac output, increases cardiac work

Adrenal Cortex
- Angiotensin-II & Angiotensin-III → increases the synthesis & release of aldosterone
- Aldosterone acts on distal tubule to promote Na^{+2} reabsorption & K^+/H^+ excretion → it occurs in very low doses

Kidney
- A-II promotes Na^+/H^+ exchange in proximal tubule → increased Na^+, Cl^- & HCO_3^- reabsorption
- it reduces renal blood flow & produces intrarenal hemodynamic effects result in Na^+ & water retention

Mechanism of action of Angiotensin receptors
- G-protein coupled receptors
- ***Vascular & Visceral smooth muscle contraction:*** Activating MLCK by phospholipase C-IP3/DAG-intracellular Ca^{+2} release
- enhanced Ca^{+2} movement induces aldosterone
- ***Myocardium, Vascular smooth muscle & fibroblasts:*** AT_1 receptors → Angiotensin-I & Angiotensin-II → activates MAP kinase, Tyrosine protein kinase, Protein Kinase C

Inhibitors of Renin-Angiotensin System
- Sympathetic blockers → decreases renin release
- Renin inhibitory peptides & renin specific antibodies block renin action → interfere with generation of A-I from angiotensionogen
- ACE Inhibitors → prevents generation of Angiotensin-II
- AT_1 receptor Antagonists → block the action of Angiotensin-II
- Aldosterone Antagonists → block mineralocortocoid receptors

Classification of Anti-Hypertensive Drugs

1. **Angiotensin Converting Enzyme Inhibitors**
- Angiotensin Converting Enzyme helps in the formation of Angiotensin-II → Vasoconstriction
- These drugs inhibits Angiotensin Converting Enzyme → inhibits the formation of Angiotensin –II → vasodilatation
    - ***Captopril, Enalapril, Lisinopril, Perindopril, Ramipril***

2. **AT₁ receptor antagonists**
- GPC Recpetor AT_1 involves in the action of Angiotensin-II & III → Vasoconstriction
- These drugs competitively antagonize AT_1 receptors → inhibits stimulation of Angiotensin-II, Angiotensin-III → Vasodilatation
    - ***Losartan, Candesartan, Irbesartan***

3. **Calcium Channel Blockers**
- By blocking calcium channels → inhibits phosphorylated myosin light chain kinase enzyme → inhibits the phosphorylisation of Myosin → not combining with actin for contraction → relaxation of peripheral vessels → decrease in peripheral resistance
    - ***Verapamil, Diltiazem, Nifedipine, Felodipine, Amlodipine, Nitrendipine, Lacidipine***

4. Diuretics
- reduces the plasma & ECF volume by decreasing the Na$^+$ & Water retention
    - *__Thiazides:__ Hydrochlorothiazide, Chlorthalidone, Indapamide*
    - *__High ceiling:__ Furosemide*
    - *__K+ Sparing:__ Spironolactone, Triamterene, Amiloride*

5. Beta Blockers
- they block the beta receptors (central sympathetic activity on heart) → they are preferred in combinations
    - *__Atenolol, Metoprolol, Propranolol__*

6. Beta + Alpha Blockers
- Reduces total peripheral resistance → better than a single beta blocker
    - *__Labetalol, Carvedilol__*

7. Alpha Blockers
- Dilates both resistance & capacitance vessels → reduction in total peripheral resistance
    - *__Prazosin, Terazosin, Doxazosin, Phentolamine, Phenoxybenzamine__*

8. Central Sympatholytics
- Decreases sympathetic outflow → fall in BP & Bradycardia
    - *__Clonidine & Methyldopa__*

9. Vasodilators
- directly arteriolar vasodilatation
    - *__Arteriolar:__ Hydralazine, Minoxidil*
    - *__Arteriolar + Venous:__ Sod. Nitroprusside*

ACE Inhibitors

- *Captopril, Enalapril, Lisinopril, Benazepril, Ramipril, Peridopril, Quinapril, Cliazapril, Zofenopril, Fosinopril*
- The first ACEI – Teprotide derived from Pit Viper Venom

Mechanism of Action
- ACE inhibitors inhibits the angitotensin converting enzyme which is helps in the formation of Angiotensin-II & III → the actions of Angiotensin – II & III are inhibited
- CVS → Vasodilatation & suppression of central sympathetic activity and also suppression of release & synthesis of adrenaline and noradrenaline → decreases Myocardial contraction by inhibiting Ca^{+2} influx, decreases Heart rate by suppressed sympathetic activity, increases Cardiac output→decreases cardiac work
- *__Adrenal Cortex:__* By inhibiting the ACE → stimulation of Angiotensin-II & Angiotensin-III are inhibited → decreases synthesis & release of aldosterone
- Due to inhibition of Aldosterone → Na^{+2} & water reabsorption is decreases → decrease in plasma volume
- By inhibiting the stimulation of Angiotensin-II & Angiotensin-III → it increases renal blood flow & result in Na$^+$ & water excretion

Captopril
- it is a sulfhydryl containing dipeptide surrogate of proline → which abolishes the pressor actions of Angiotensin-I
- It increases plasma kinin levels & potentiate the hypotensive action of exogenously administered bradykinin
- Elevated Kinins are responsible for cough & angioedema induced by ACE inhibitors

Adverse Effects
- *Hypotension:* decreases BP in diuretic treated & CHF patients
- *Hyperkalemia:* risk in patients with impaired renal function & in those taking K+ sparing diuretics, NSAIDs or β-blockers
- *Cough:* occurs within 1-8 weeks and subsides within 4-6 after discontinuation
- *Rashes, Urticaria*

- *Angioedema:* swelling of lips, mouth, nose, larynx, airway obstruction
- *Dysguesia:* loss or alteration of taste
- *Fetopathic:* fetal growth retardation, hypoplasia of organs & fetal death → ACE inhibitors prescribed in later half of pregnancy
- *Headache, dizziness, nausea, bowel upset*
- *Granulocytopenia & proteinurea*
- *Acute renal failure*

Enlapril

Advantages over Captopril
- more potent, effective dose 5-20 mg OD or BD
- Its absorption is not affected by food
- Onset of action is slower (due to need for conversion to active metabolites), less liable to cause abrupt first dose hypotension
- Longer duration of action → can be used in all types
- Rashes & dysguesia is rare

ACE Inhibitors - Uses

- **Hypertension:** they are mostly used in renovascular & resistant hypotension
- they are first line drugs in all types of HTN

advantages over other drugs as antihypertensives
  - lack of postural hypotension, electrolyte disturbances, feeling of weakness, CNS effects
  - Safety in asthmatics, diabetics, peripheral vascular disease
  - prevention of secondary hyperaldosteronism & K+ loss due to diuretics
  - renal blood flow is well maintained
  - It reverses left ventricular hypertrophy & increased wall-to-lumen ration of blood
  - No hyperuricaemia, no deleterious effect on plasma lipid profile
  - No rebound hypertension on withdrawal
  - it doesn't disturb day-to-day life quality like work performance, sleep, sexual performance
- **CHF:** cause both arteriolar & venodilatation in CHF patients: reduce after load & preload
- decreases in right atrial pressure, pulmonary arterial pressure, pulmonary capillary wedge pressure, systemic vascular resistance, systolic wall stress & systemic BP
- Decreased Heart Rate, increased Cardiac Output →decreased Cardiac work → exercise capacity is enhanced
- Accumulated salt & water are lost due to improved renal perfusion & abolition of mineralocortcoid mediated Na^+ retention
- prolonged survival of CHF patients of all grades
- They reduce episodes of decompensation, MI & sudden death
- **MI:** These drugs given within 24 hrs of an attack and continued for 6 weeks → reduce early as well as long term mortality
- **Prophylaxis in High Cardiovascular risk subjects:** prevents risk of developing heart failure & MI
- **Diabetic nephropathy:** used to prevent end stage renal disease in type-1 & type-2 DM
- **Scleroderma crisis:** marked rise in BP & deterioration of renal function

Angiotensin Antagonists

- Alternatives to ACEI
- *Losartan, Candesartan, Irbesartan, Telmisartan, Eprosartan*
- *Losartan:* competitive antagonist of A-II
- more selective for AT_1 than AT_2
- it blocks all actions of A-II like vasoconstriction, central & peripheral sympathetic stimulation, release of aldosterone & Adr from adrenals, renal actions promoting salt & water reabsorption,

central actions like thirst, vasopressin release & growth promoting actions on heart & blood vessels

AT₁ antagonists VS ACE inhibitors

- AT$_1$ don't interfere with degradation of bradykinin & other ACE substrates → no rise in bradykinin level
- complete inhibition of AT$_1$ receptor activation
- Alternative pathways of A-II generation & AT$_1$ receptor activation with ACE inhibitors
- Indirect AT$_2$ activation due to more blockade of AT$_1$ where ACE inhibitors depress both AT$_1$ & AT$_2$
- Free from cough & dysguesia

Adverse Effects
- cause hypotension & hyperkalemia
- headache, dizziness, weakness & upper g.i. side effects are mild
- Losartan has fetopathic potential → so avoid in preganancy

Calcium Channel Blockers

- these are also 1st line antihypertensive drugs
- They act by decreasing peripheral resistant

advantages of CCBs as antihypertensives
- no impairment of physical work capacity
- no sedation
- can be used in asthma, angina
- don't impair renal perfusion
- don't affect male sexual function
- no effects on plasma lipid profile, uric acid level & electrolyte balance
- minimal effect on quality of life
- can be used in pregnancy (not before labour)

disadvantages of CCBs as antihypertensives
- negative inotropic/dromotropic action may worsen CHF & cardiac conduction defects
- Nifedipine decreases insulin release
- CCBs may accentuate bladder voiding difficulty in elderly males & gastroesophageal reflex

Diuretics - Thiazides

- the diuretic of choice in uncomplicated hypertension
- They inhibit Na^+ - Cl^- symport
- They reduces plasma & ECF volume by about 15% → decreased CO
- Fall in BP is maintained by slowly developing reduction in TPR
- Decreased intracellular Na^+ in vascular smooth muscle → decreased stiffness of vessel wall, increases compliance & no contraction
- no effect on heart rate, Cardiac output, capacitance vessels, sympathetic reflexes
- Postural hypotension is rare
- They potentiate all other anti-HTN drugs except DHPs

Diuretics – High Ceiling

- they are weak anti-HTN than thiazides
- fall in BP is entirely dependent on reduction in plasma volume & cardiac output
- They cause fluid & electrolyte imbalance, weakness & other side effects
- they are used only when thiazides are complicated
    - chronic renal failure
    - Coexisting refractory CHF

- Resistance to combination

Diuretics - disadvantages

- *Hypokalemia* – muscle pain, fatigue, loss of energy & impotence
- *Carbohydrate Intolerance:* inhibition of insulin → precipitates diabetes
- *Dyslipidemia:* rise in total & LDL cholesterol & Triglycerides with lowering HDL → can increase atherogenic risk
- *Hyperuricaemia:* inhibits urate excretion → incidence of gout
- Dose dependent *increased incidence of sudden cardiac death*

β+α - Blockers

- *Labetalol:* reduces TPR & acts faster than pure β-blockers
- it has been used i.v. for rapid BP reduction in cheese reaction, clonidine withdrawal
- *Carvedilol:* non-selective β + selective $α_1$ blocker → produces vasodilatation
- it is also used in CHF

Prazosin

- Prototype $α_1$ receptor antagonists
- It causes peripheral vasodilatation & fall in arterial pressure with lesser tachycardia → because
- It lacks $α_2$ receptor blocking actions actions → doesn't promote NE release from sympathetic nerve terminals
- It decreases cardiac preload
- Partly because it suppresses sympathetic outflow from CNS
- It is also a potent inhibitor of cyclic phosphodiesterase enzyme → rise in cAMP in vascular smooth muscle → contribute to its vasodilating effect
- Another: rise in concentrations of HDL & decrease in LDL & triglycerides
- Selective $α_1$ agonist → it relaxes → smooth muscle in the bladder neck, prostate capsule & prostatic urethra → improves urine flow in BHP

Therapeutic uses

- **Treatment of Hypertension:** α1 receptor antagonists improve lipid profile
- **Dose :** 1mg given at bed time to reduce the risk of postural hypotension → then dose is titrated upward depending on BP
- **To treat BHP :** Dose: 1-5 mg twice daily
- **In Raynaud's disease:** it decreases the incidence of digital vasospasm [but Ca+2 channel blockers are more preferred]
- **In patients with Mitral and Aortic Valvular insufficiency:** it decreases cardiac afterload

Adverse Effects

- Marked Postural Hypotension at initial dose
- Impotence, nasal decongestion, GIT upset & Na+ & water retention

Advantages of Prazosin

- Improves carbohydrate metabolism → suitable for diabetics
- favorable effect on lipid profile: lowers LDL cholesterol & triglycerides, increases HDL
- no impairment of cardiac contractility
- Doesn't alter exercise capacity & uric acid levels
- also used in BHP

Central Sympatholytics

Clonidine

- $α_2$ agonist drug which is used in the treatment of Hypertension
- After IV injection (higher plasma concentrations) → produces a transient rise in BP , Oral administration (low plasma concentrations) → fall in BP

- It has lesser intrinsic activity on α2 receptors present postsynaptically on vascular smooth muscle, at lower plasma concentrations

Basis of antihypertensive action
- Mainly stimulates α2 receptors present at vasomotor center → central sympathetic flow is reduced → fall in BP & HR
- It binds and stimulates the imidazoline receptors in brain (I_1) → modulate central α2 activity → reduces central sympathetic outflow
- Activates presynaptic α2 receptors present on sympathetic postganglionic neurons → suppress the release of Noradrenaline from peripheral nerve endings
- Decreased central sympathetic flow → reduces rennin release & decreases renal vascular resistance
- Its is especially used in the treatment of Hypertension with poor renal functions
- After Oral administration, bioavailabilty is 100%

Therapeutic uses
- **Moderate Hypertension:** orally 100μg BD along with a diuretic (to counteract the sodium & water retention which leads to the development of tolerance)
- **To control diarrhea in diabetic patients with autonomic neuropathy:** stimulation of α2 receptors in GIT increases Na+ & water retention & inhibits HCO3- secretion → reduces intestinal motility by inhibiting the release of Ach
- **In prophylaxis Migraine:** it reduces cerebral blood flow (25-75μg)
- **Management of Nicotine, Alcohol & Opiate withdrawal**
- **Preanesthetic medication:** its used preoperatively for its sedative, anxiolytic & analgesic effects → to provide heamodynamic stability & to reduce anaesthtic requirements
- It attenuates perioperative stress → reduce perioperative oxygen consumption
- **Menopausal hot flushes**
- **For diagnosis of Phaeochromocytoma**
- **For relief of Pain:** epidurally along with opioids in postoperative, neuropathic & labour pain

Adverse effects: Rebound hypertension, Dry mouth, Seadation, Nasal Stuffiness, Constipation, Impotence, Contact dermatitis

Vasodilators

Hydralazine

- directly acting arteriolar vasodilator with little action on venous capacitance vessels
- Reduces TPR
- It causes greater reduction of diastolic than systolic blood pressure
- It causes tachycardia, increase in cardiac output & renin release → increase aldosterone → Na+ & water retention
- Direct augmentation of NA release → myocardial contractility
- hyperdynamic circulatory state is induced → ppt angina due to increased cardiac work or steal phenomenon
- No reduction in renal blood flow

Adverse Effects
- they are mostly due to vasodilatation
- Facial flushing, conjuctival injection, throbbing headache, diaainess, palpitation, nasal stuffiness, fluid retention, edema, CHF
- angina & MI may presipitate inpatients with coronary artery disease
- paresthesias, tremor, muscle cramps, edema, rarely peripheral neuritis
- On prolonged use → lupus erythematous or rheumatoid arthritis

Aldosterone Antagonists

Spironolactone , Eplerenone

- they are majorly used in resistant hypertension
- they are effective in all types of hypertension regardless of renin level
- Aldosterone plays a central role in target organ damage, including the development of ventricular & vascular hypertrophy & renal fibrosis → antagonists improves these conditions
- Spironolactone – breast pain, gynacomastia, hyperkalemia
- Treatment of Hypertension
- The main aim of the treatment is to prevent morbidity & mortality with minimum inconvenience to the patient

Complications associated with Hypertension

- Cerebrovascular disease, Transient ischaemic attacks, stroke
- Hypertensive heart disease – left ventricular hypertrophy, CHF
- Coronary artery disease, angina, myocardial infarction, sudden cardiac death
- Arteriosclerotic peripheral vascular disease, retinopathy
- Dissecting aneurysm
- Glomerulopathy, renal failure

WHO-ISH guidelines

- Except for grade III HTN, start with a single most appropriate drug
- Initiate therapy at low dose; thiazide dose should not be > 12.5-25 mg/day
- If only partial response, add a drug from another complimentary class or change to low dose combination
- If no response, change to a drug from another class
- One should be Thiazide in combination

Combination Therapy

- Drugs which increase plasma renin activity – diuretics, vasodilators, CCBs, ACEI may be combined with drugs which lower plasma renin activity – β blockers, clonidine, methyldopa
- All sympathetic inhibitors (except β blockers) & vasodilators cause fluid retention: used alone tolerance develops → diuretic prevents fluid retention & tolerance development
- Hydralazine & DHPs cause tachycardia which is counteracted by β blockers
- Increase in TPR due to initial doses of β blockers is antagonized by vasodilator
- ACEI/AT1 antagonists are synergistic with diuretics → preferred combination for patients with CHF & left ventricular hypertrophy
- ACEI+CCBs ; β blocker or Clonidine or Methyldopa ; β blocker + prazosin

Treatment of Hypertension in Pregnancy

Drugs to be avoided

- Diuretics: reduce blood volume – accentuate uteroplacental perfusion deficit – increase risk of foetal wastage, placental infarcts, miscarriage, stillbirth
- ACEI,AT1 antagonists: risk of foetal damage, growth retardation
- Reserpine: suicidal depression in mother; nasal obstruction,deranged respiratory & temperature control in new born
- Non-selective β blockers: Propranolol causes low birth weight, decreased placental size, neonatal bradycardia & hypoglycaemia
- Sod nitroprusside: contraindicated in eclampsia

Safer Drugs

- Hydralazine, Methyldopa
- DHP CCBs: discontinue before labour → as they weaken uterine contractions
- Cardioselective β blockers & with ISA(intrinsic sympathomimetic activity) – atenolol, pindolol, acebutolol
- Prazocin & clonidine – provided that postural hypotension can be avoided
- Hypertensive Emergencies → reduction of blood pressure within 1 hr to avoid the risk of serious morbidity or death →labetolol or nicardipine
- Hypertensive Urgencies → blood pressure must be reduced within a few hours

Congestive Heart Failure

- Cardiac output is insufficient to meet the demands of tissue perfusion
- HF may be due to systolic dysfunction or diastolic dysfunction
- *Systolic dysfunction:* Ventricles are dilated & unable to develop sufficient wall tension to eject adequate quantity of blood ← Ischaemic heart disease, valvular incompetence, dilated cardiomyopathy, myocarditis, tachyarrhythmias
- *Diastolic dysfunction:* the ventricular wall is thickened & unable to relax properly during diastole; ventricular filling is impaired because of which output is low ← in sustained hypertension, aortic stenosis, congenital heart disease, A-V shunts, hypertrophic, cardiomyopathy

Classification

Drugs with positive inotropic effects

1. *Cardiac Glycosides:* Digoxin, Digitoxin & Ouabain
2. *Bipyridines or Phosphodiesterase Inhibitors:* Inamrinone, Milrinone & Levosimendan
3. *Beta-agonists:* Dopamine, Dobutamne & Dopexamine

Drugs without positive inotropic effects

1. *Diuretics:* Bumetanide, Furosemide, Hydrochlorothiazide, Metolazone & Spironolactone
2. *ACEIs:* Enalapril, Lisinopril, Ramipril
3. *Beta$_1$-Blocker:* Bisoprolol, Carvedilol, Metoprolol
4. *Vasodilators:* Hydralazine, Sodium nitropruside, Isosorbide Dinitrate, Nesiritide

Treatment of CHF

Goal of the therapy

❖ *Relief of congestive/low output symptoms & restoration of cardiac performance:*
✓ Inotropic Drugs → Digoxin, dobutamine/dopamine, amrinone/milrinone
✓ Diuretics → Furosemide, thiazides
✓ Vasodilators → ACE inhibitors/AT$_1$ antagonists, Hydralazine, nitrate
✓ β-blocker→ Metoprolol, Bisoprolol, Carvedilol
❖ *Arrest / reversal of disease progression & prolongation of survival*
✓ ACE inhibitors/AT$_1$ antagonists
✓ β-blockers
✓ Aldosterone antagonist
✓ Rest & salt restriction

Cardiac Glycosides

- Sugar molecule is joined together with a non-sugar molecule by a ether it is called a glycoside → on hydrolysis → sugar & non-sugar part
- Non-sugar part or aglycone → pharmacological activity

- Sugar part → guides pharmacokinetic profiles like lipid solubility & Cell permeability
- *Digitalis purpurea* is major source of digitoxin
- **Mechanism of Action**
- **_Normal Ionic movements during the contraction of cardiac mucle_**
- the force of contraction of the cardiac muscle is directly related to the conc of free cystolic Ca^{+2}
- any drug which can increase cystolic Ca_{+2} → can increase the FC of heart muscle & also increase sensitivity of the contractile machinery to Ca^{+2}
- Initially Ca^{+2} entry through sensitive L-type Ca^{+2} channels → triggers the release of a Ca^{+2} from sarcoplasmic reticulum → by activating SR-Ca^{+2} channel
- increased Ca^{+2} conc → initiates the contractile process
- during restorative process, or periodic contraction of cardiac muscle, Ca^{+2} ions are removed by reuptake into the sarcoplasmic reticulum by SR-Ca^{+2} ATPase & by its expulsion from the cell by a Ca^{+2}/Na^+ exchange pump
- Intracellular $Na+$ balance is restored by Na^+/K^+ ATPase

- **_Cellular Mechanism of Digitalis Action_**
- Digitalis binds to & reversibly inhibits cardiac cell membrane associated Na^+/K^+ ATPase → leads to progressive accumulation of intracellular Na^+ & loss of intracellular K^+
- increase in the intracellular $Na+$ conc → prompts the influx of Ca^{+2} from the outside & outflux of $Na+$ from inside by Ca^+/Na^+ exchange mechanism
- Na^+/Ca^{+2} exchanger normally → extrudes Ca^{+2} in exchange for $Na+$
- in the presence of increased intracellular $Na+$ conc → it extrudes Na^+ in exchange of extracellular Ca^{+2}
- an increase in Ca^{+2} permeability through voltage sensitive L-type $Ca+2$ channels during plateau phase → increased cytosolic free $Ca+2$ by these two mechanisms → triggers the release of more $Ca+2$ from sarcoplasmic reticulum & mitochondria
- Digitalis also inhibits SR-Ca^{+2} ATPase & reduces the reuptake of Ca^{+2} by sarcoplasmic reticulum
- the inotropic action of digitalis results due to an increase in cytoplasmic $Ca+2$ conc that enhances cardiac muscle contractility & cardiac output
- Higher serum K^+ concentration inhibits digitalis binding to Na^+/K^+ ATPase → hyperkalaemia reduce digitalis toxicity & Hypokalaemia increase the risk of toxicity
- Hypercalacemia or hypomagnesaemia increase the risk of digitalis induce arrhythmias

- **Pharmacological Actions**
- **_Cardiac Effects of Digitalis_**
- effects of digitalis on normal heart are different than on the failing heart
- Normal Heart: Digitalis → increased Force of Contraction, increased Cardiac Output, cause constriction of peripheral blood vessels
- Heart Failure: increased contractility → increased Cardiac Output
- Cardiac systole is shortened → more time is provided for ventricular rest & filling
- HR is reduced due to direct (Na^+-K^+ ATPase inhibition) & direct (vagal stimulation) effects on SA node
- As heart improves → more O_2 needed due to increase in force of contraction → it is nuetralized by reduction in heart rate & reduction in ventricular size → reduce the O_2 requirement of the heart
- Digitalis decreases the conduction velocity of AV node & His-Purkinje system & prolongs their effective refractory period by both vagal & extra-vagal (Na^+-K^+ ATPase inhibition) action
- Increased ERP & decreased conductivity of AV node protects ventricles from atrial flutter or atrial fibrillation

- In smaller doses → digitalis increases the conduction velocity & decreases the ERP of atrial muscle via vagal action
- In higher doses → increases automaticity & contractility but shortens ERP of atrial muscle & ventricles causing extrasystoles, pulsus bigeminus, ventricular fibrillation
- ***Effects on ECG***
- Prolongation of PR interval (delayed AV conduction)
- Shortening of QT interval (shorter ventricular systole),
- depression of ST segment & inversion or disappearance of T-wave
- ***Extra-Cardiac Effects of Digitalis***
- ***Blood Vessels:*** In normal persons → vasoconstrictor effects
- **In heart failure** → digitalis opposes sympathetic over activity & causes decreased Heart Rate, decreased Peripheral vascular Resistance (decreased in afterload), increased venous tone (decreased preload)
- Leeser blood reaches the heart, end diastolic volume is reduced
- No effect on BP
- ***Kidney:*** more diuresis due to more renal perfusion → shift of oedematous fluid into ciculation
- other reasons → decreased in compensatory sympathetic discharge, increased in activation of RAAS (decreased renin, decreased AT-II, decreased aldosterone release, decreased salt & water retention)
- ***GIT:*** anorexia, diarrhoea, nausea & vomiting
- ***CNS:*** disorientation, hallucinations, visual disturbances & aberrations of color perception

- **Therapeutic Uses**
- ***Congestive Heart Failure***
- it is a drug of choice for "low output heart failure" due to HT, IHD or arrhythmias
- It is not suitable for "high output heart failure" due to thyrotoxicosis or arteriovenous shunt
- it relief from dyspnoea & cyanosis by subsiding pulmonary congestion
- it increases urinary output & reduces oedematous fluid by increasing renal perfusion
- with decreased end diastolic fibre tension, heart size & O_2 demand of myocardium decreases
- Bradycardia & improved myocardial circulation with the development of a sense of well being
- ***PSVT***
- it is a common arrhythmia due to reentry phenomenon taking place at SA or AV node
- They respond to digitalis → because of reflex vagal activation which slows the conduction of impulses
- Adenosine or Verapamil or β-blockers are more preferred → more effective
- Digitalis is reserved for preventing recurrences
- ***Atrial Flutter & Atrial Fibrillation***
- Atrial flutter (rate 200-300beats/min) & atrial fibrillation (rate 500 beats/min) → are effectively treated with digitalis → as it decreases conduction velocity & increases ERP of AV node
- If the ventricular rate is 72-80 beats /min → add β blocker or verapamil
- ***T/t of Dilated Heart***
- Digitalis is the most preferred drug for patients having dilated heart & low ejection fraction → as it helpful in restoring cardiac compensation
- It is usually prescribed when the patient's condition is not controlled by diuretic or ACEI

- **Adverse Effects**
- ***Cardiac side effects***
- bradycardia, partial or complete heart block, atrial or ventricular extrasystoles, coupled beats, ventricular fibrillation & fatal cardiac arrhythmias

- *Extra-Cardiac side effects*
- *GIT:* Anorexia, nausea, vomiting, diarrhoea, abdominal cramps
- *CNS:* headache, fatigue, neuralgia, blurred vision, loss of color perception
- *Endocrinal:* Gyanecomastia (in males)-rare
- **Digitalization**
- *Slow*
- 0.25 mg/day, if adequate response is not there then increase to 0.375 and to 0.50 mg
- *Rapid*
- 0.5-1.0 mg stat followed by 0.25 mg every 6 hours with careful monitorning
- *Emergent*
- 0.25 mg followed by 0.1mg hourly given by slow iv infusion
- **Phosphodiesterase Inhibitors**
- *Inamrinone, Milrinone, Levosimendan*
- they are non-glycoside, non-sympathomimetic inotropic agents
- They inhibit Phosphodiesterase isoenzyme-3
- This enzyme is located in cardiac myocytes & vascular smooth muscle → helps in the conversion of cAMP to AMP
- these drugs inhibit PDE and increase the levles of cAMP → increased myocardial contractility along with vasodilatation
- Inamrinone → causes nausea, vomiting, dose dependent thrombocytopenia, arrhythmias
- Milrinone → safer but also potentiate arrhythmias

- **β-Agonists**
- *Dobutamine, Dopamine*
- β-Agonists improves cardiac performance through positive inotropic effects
- B1 agonists lead to an increase in intracellular cAMP which results in the activation of protein kinase
- Slow Ca^{+2} channels are then phosphorylated by protein kinase which increases ca^{+2} flow into the myocardial cell causing → increased force of contraction of heart muscle
- **Dobutamine**
- most commonly used inotropic agent other than digitalis
- It has more selective inotropic effects
- More selective β_1 agonist
- $T_{1/2}$ – 2 min → should be given by iv infusion
- Css is achieved in 10 min
- Used for short term management of acute HF & for patients of HF associated with MI or cardiac surgery
- It increase Cardiac Output, urinary output, stroke volume without affecting Heart Rate, Total peripheral resistance, Blood pressure
- It also increases O_2 deman → angina may precipitate or MI may aggravated
- Tolerance will be developed
- It increases AV conduction→ caution while using in atrial fibrillation
- It increases BP → avoid in HTN
- **Dopamine**
- In low doses, it acts on D_1 receptors in kidney → causes dilatation of renal blood vessels resulting in an increase in GFR, renal blood flow & urinary output
- In therapeutic doses, it stimulates β_1 receptors resulting in an increase in cardiac output but total peripheral resistance unchanged, due to splanchnic vasodilatation through stimulation of D_1 receptors
- In higher doses, it activates α_1 receptors → causes vasoconstriction with an increase in TPR & pulmonary pressure → this nullifies the beneficial effects of dopamine in heart failure

- Dopamine is used in low cardiac output heart failure with compromised renal functions (advantages of increased CO with renal vasodilatation)
- Diuretics
- Loop diuretics like Bumetanide & Furosemide → promptly reduce pulmonary oedema due to rapid diuresis→ results in decreased Extra Cellular Fluid, decreased venous return→ decreased right ventricular output → it results reduction in oedema & reduction in cardiac size which leads to an improvement in pump efficiency
- Due to more loss of H^+ & K^+ → leads to arrhythmias & enhance digitalis toxicity → can be overcome by addition of an Aldosterone antagonist

- **Spironolactone+digitalis+ACEI in CHF**
- combination improves survival among patients with chronic/severe HF
- Spironolactone → enhances diuresis by promoting Na+ & water excretion → prevents myocardial & vascular fibrosis
- It also helps in restoration of the diuretic response to loop diuretics when refractoriness to them develops on chronic use
- Hyperkalemia is the major risk during the therapy

- **ACEIs & ARAs**
- They cause both arteriolar & venodilatation in CHF patients: reduce after load & preload
- Decreases in right atrial pressure, pulmonary arterial pressure, pulmonary capillary wedge pressure, systemic vascular resistance, systolic wall stress & systemic BP
- Decreases Heart Rate, increases Cardiac Output →decreased Cardiac work → exercise capacity is enhanced
- Accumulated salt & water are lost due to improved renal perfusion & abolition of mineralocortcoid mediated Na^+ retention
- prolonged survival of CHF patients of all grades
- They reduce episodes of decompensation, MI & sudden death

- **β_1- Blockers**
- They are contraindicated in CHF
- Heart Failure→ decreased cardiac output ← increased heart rate should be there to maintain, but β-blockers decreased Heart Rate, decreased cardiac contractility → produces acute cardiac decompensation
- But, Bisoprolol, Carvedilol, Metoprolol → improves ventricular functions & prolong the survival
- Vasodilators
- They reduce pulmonary congestion & increases Cardiac output by reducing preload and/or afterload
- Rationale Bisoprolol in CHF
- Beta blockers reduce myocardial contractility & apparently they are contraindicated in CHF
- But Bisoprolol & Carvedilol are most widely used in CHF
- They decrease the rate of mortality in CHF patients
- ❖ **They cause**
    - i. Bradycardia → increases coronary flow
    - ii. Reduction of rennin activity
    - iii. Protection from catecholamine induced injury to myocardium
    - iv. Prevention of myocardial cell apoptosis & ventricular remodeling

Treatment of CHF with Pulmonary Edema

❖ Morphine 2-8mg iv and repeated after 2-4 hrs → Morphine increases venous capacitance, lowering LA pressere and relieves anxiety → which can reduce the efficiency of ventilation
❖ Intravenous Furosemide 40 mg → venodilatation
❖ Sublingual GTN or Isosorbide dintrate → they accelerates clinical improvement by reducing both BP & LV filling pressures
❖ Nitrates will ameliorate dyspnea rapidly prior to the onset of dieresis
❖ Inhaled β-agonists or Intravenous aminophylline → may reduce the bronchospasm

Antiarrhythmic Drugs

❖ Cardiac Arrhythmias → deviations from the normal pattern of cardiac rhythm
❖ Include abnormalities in impulse formation → Heart rate, its regularity, site of impulse origin, disturbances in impulse conduction which disrupt the normal sequence of atrial & ventricular depolarization
❖ Cardiac Arrhythmias → too slow or too rapid or asynchronous → reduce cardiac output & at times may precipitate ventricular fibrillation.
❖ Some times these drugs even precipitate lethal arrhythmias → for the treatment of asymptomatic arrhythmias these drugs are avoided.

Cardiac Electrophysiology

❖ **SA Node:** Small collection of cells that initiate atrial systole

❖ **AV Node:** 1. Provides delay in impulse transmission
  2. Protects ventricles from atrial fibrillation

❖ **Bundle of His:** 1. Divides R and L bundles
  2. Provides orderly depolarization of ventricles

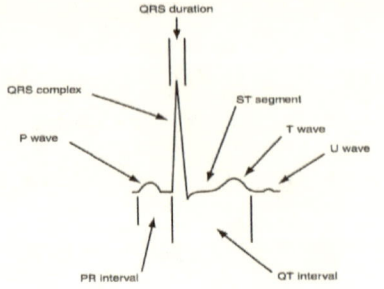

P wave → Atrial Systole
PR interval → delay at AV node
QRS complex → Ventricular systole
T wave → Ventricular repolarization

ECG

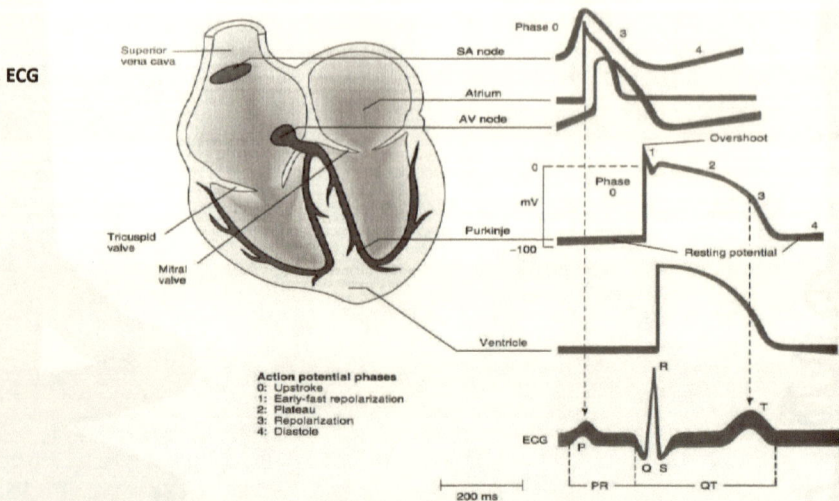

Action potential phases
0: Upstroke
1: Early-fast repolarization
2: Plateau
3: Repolarization
4: Diastole

200 ms

1. Impulse Generation and Transmission

Two types of Cardiac tissue

    a)   ***Pacemaker or Conductive tissue*** → SA node, AV node, bundle of His & Purkinje fibres.
          (SA node over-rides other pacemaker cells)
    b)   ***Myocardial contractile tissue*** → which includes atria & ventricles

❖   Two electrical sequences make the heart chambers to fill with blood and then contract to supply it to various organs
❖   The first electrical sequence is the "Impulse formation" & this takes place when an electrical impulse is generated spontaneously & automatically.
❖   The second sequence is "Impulse transmission"
❖   The normal pathway of impulse conduction is as follows

Atrial Muscle

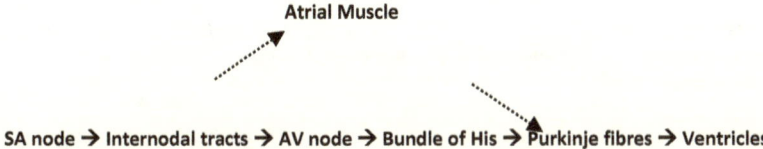

SA node → Internodal tracts → AV node → Bundle of His → Purkinje fibres → Ventricles

❖   SA node in the wall of right atrium, the main pace maker and initiates a spontaneous action potential.
❖   Impulse generated by SA node → trigger Atrial contraction → impulse travels through Internodal tracts to AV node (situated at the lower interatrial septum) → at this site, the impulses are delayed to permit completion of atrial contraction.
❖   Impulse reaches the Bundle of His → it travels along the right & left branches located on either side of the intraventricular septum → from here impulse reaches to Purkinje fibres → ending in the Ventricular endocardial surfaces → Ventricular contraction

2. Myocardial Action Potential

❖   For Cardiac contractions → cardiac cells must depolarize & then repolarise through alterations in their ionic permeabilities (mainly of Na^+, K^+, Ca^{+2}) across the cell membrane

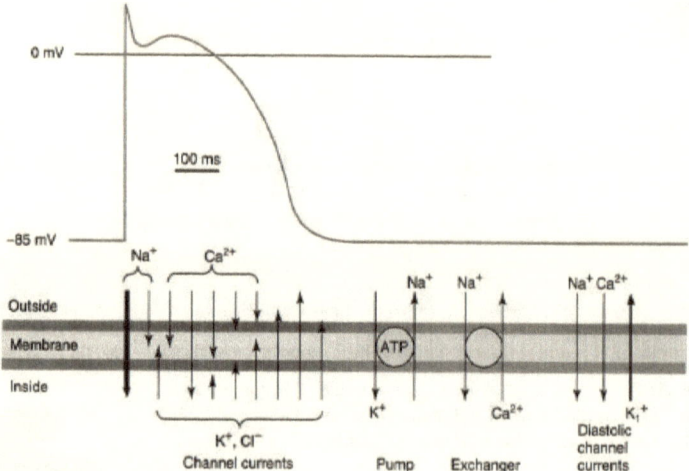

❖ The Action Potential can be divided into five phases: *Rapid depolarization* (Phase 0), *Early repolarization* (Phase 1), *Plateau* (Phase 2), *Rapid repolarization* (Phase 3), *Slow diastolic depolarization* (Phase 4)

Phase 0 → Na⁺ enters the cell through fast inward Na-channels → cell membrane's electrical charge from negative to less negative & finally positive → conformation of Na⁺ channels changes → inactivated

Phase 1 → Voltage gated Na⁺ channels close & K⁺ channels rapidly open → rapidly repolarises → produces spike & dome formation in graph → short-lived outward current due to opening of both K⁺ channel & Cl⁻ channels

Phase 2 → Voltage sensitive L-type Ca+2 channels open → causing slow inward current → this is balanced by slow outward K+ current i.e., depolarization = repolarization → AP reaches Plateau

Phase 3 → Ca+2 channels close & K+ channels open → K+ are pumped out of the cell → cell rapidly repolarises → resumes its initial negativity → from Phase 0 to Phase 3, there has been a gain of Na+ & loss of K+ → This imbalance is corrected by Na+-K+ ATPase pump

Phase 4 → electrically stable → Na+ channels involved in atrial & ventricular myocyte depolarization recover completely from inactivation→ available for activation again

Action Potential in Nodal Tissues

❖ Ca+2 influx dominates the upstroke (depolarization) & largely responsible for initiation & propagation of AP → cardiac automaticity is decreased by Ca+2 channel blockers (*Automaticity means the ability of a cell to alter its resting membrane potential toward the excitation threshold potential without the influence of any stimulus*)

❖ SA node over rides the other cells which have the potential automaticity.

❖ The other cells can become pacemakers when their own intrinsic rate of depolarization becomes greater than that of SA node or when the pacemaker cells within the SA node are depressed.

❖ AV node & Purkinje fibres are only Latent Pacemakers.

❖ The rate of pacemaker discharge within these cells is influenced by Sympathetic & Parasympathetic activity.

❖ **Stimulation of β₁ receptors** → makes the slope of diastolic depolarization (Phase 4) more steeper → increase the pacemaker activity

❖ **Stimulation of M2 receptor on SA node** → decreases the diastolic depolarization

❖ **Release of ACh from Cholinergic vagal fibres** → increases the outward K+ conductance in pacemaker cells → results in more negative potential or hyperpolarisation

Genesis of Cardiac Arrhythmias

The abnormalities in cardiac rhythm & in the conduction of cardiac impulse → result in **Bradyarrhythmias or Tachyarrhythmias**

Bradyarrhythmias (heart rate is low < 60 beats/min)

❖ Result from the failure of the impulse generation within the SA node or failure of impulse conduction through the SA node

❖ They include various kinds of heart block complete cessation of electrical activity

❖ Treatment: permanent cardiac pacemakers

Tachyarrhythmias (heart rate is high > 100 beats/min)

❖ These are can be cured with the help of drugs

❖ These include Atrial Fibrillation (HR is completely irregular), Supraventricular Tachycardia (HR is rapid but regular), Sustained Ventricular Tachyarrhythmias

❖ Sustained Ventricular Tachyarrhythmias are less common but serious. (they include Ventricular Tachycardia, Extra systoles & Ventricular Fibrillation)

The genesis of cardiac arrhythmias involves abnormal impulse generation, triggered activity or abnormal impulse conduction.

Abnormal Impulse Generation

❖ Depressed automaticity of SA node → leading to escape beats and bradycardia.

❖ Enhanced automaticity of SA node → leading to Sinus tachycardia.

- Av node, Purkinje fibres, atrial & ventricular cells → show spontaneous diastolic depolarization & repetitive firing → when membrane potential comes to -60 mV → this leads to ectopic beats
- Ectopic beats → stimulated by a) faster phase 4 depolarization due to ischaemia, digitalis, catecholamines, acidosis, hypokalemia or stretching of cardiac cells b) less negative resting membrane potential or c) more negative threshold potential due to ischaemia

Triggered activity

- Self sustained and depend on the preceding AP.
- Two major forms "Delayed after depolarization" & Early after depolarization

Delayed After-Depolarization (DAD)

- Due to increased intracellular Ca+2 ions concentration (Digitalis intoxication → a normal AP followed by DAD)

- DADs interrupt phase 4 of normal cardiac AP.
- These are exacerbated by fast heart rates & responsible for arrhythmias from digitalis toxicity,

catecholamines or myocardial ischaemia.

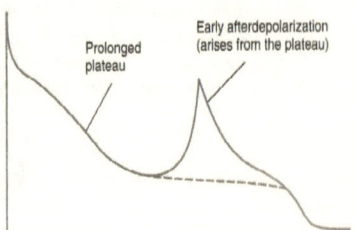

Calcium overload

Delayed afterdepolarization (arises from the resting potential!)

Early After-Depolarization (EAD)

- Phase 3 (repolarization) of normal cardiac potential is interrupted.

- Prolong cardiac repolarization phase → so these are associated with the development of polymorphic

ventricular tachycardia with a long QT interval →

known as *Torsades De Pointes Syndrome*

- Quinidine, Sotalol, Dofetilide & conditions leading to bradycardia → leads to EAD → facilitate development of

Prolonged plateau

Early afterdepolarization (arises from the plateau)

Torsades De Pointes Tachyarrhythmias.

Abnormal Impulse Conduction

- ❖ Conduction Block
- ❖ Reentry Phenomenon (Circus movement of rhythm)

Conduction Block

- ❖ Incomplete or complete block of conduction can lead to partial heart block (dropped beats) or complete heart block (idioventricular rhythm)
- ❖ Depressed conduction may result in AV nodal block or His bundle branch block

Reentry Phenomenon (Circus movement of rhythm)

- ❖ 80-90% arrhythmias occurs due to this
- ❖ Underlying mechanism for premature beats, paroxysmal Supraventricular tachycardia, atrial flutter & ventricular fibrillation
- ❖ Bidirectional conduction → occur because the reentrant impulse will reach the normal cells earlier while they are still in a refractory phase → Wolf Parkinson White Syndrome

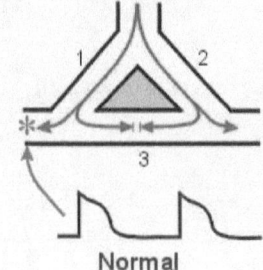

Normal

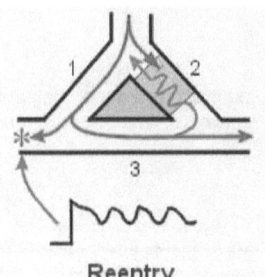

Reentry

Possible Mechanism of Antiarrhythmic Drug Action

- ❖ Suppressing the enhanced automaticity of pacemaker or non-pacemaker tissues by
  - a) Decreasing the rate of phase 0 depolarization
  - b) Decreasing the slope of phase 0 depolarization
  - c) Increasing the duration of effective refractory Period
  - d) Making resting membrane potential more negative
  - e) Making threshold potential less negative
- ❖ Abolishing reentry by further slowing the impulse conduction & thus converting the unidirectional block to bidirectional block
- ❖ They also prevent reentry by increasing the ERP of cardiac fibres surrounding the region of reentrant circuit

Antiarrhythmic agents are used to suppress fast rhythms of the Heart (cardiac arrhythmias), such as

❖ Atrial Fibrillation,
❖ Atrial Flutter,
❖ Ventricular Tachycardia
❖ Ventricular Fibrillation

Vaughan Williams Antiarrhythmic Classification

- **Class I agents interfere with the sodium (Na+) channel.**

  ➢ **Class IA :** *Quinidine, Procainamide & Disopyramide*

  ➢ **Class IB :** *Lidocaine, Phenytoin, Mexiletine & Tocainide*

  ➢ **Class IC :** *Flecainide, Encainide, Propafenone & Moricizine*

- **Class II agents are anti-sympathetic nervous system agents (beta blockers)**
  Propranolol, Acebutalol, Esmolol

- **Class III agents affect potassium (K+) efflux.**
  Amiodarone, Bretylium, Ibutilide, Dofetilide, Sotalol

- **Class IV agents affect calcium channels and the AV node.**
  Verapamil, Diltiazem

- **Class V agents work by other or unknown mechanisms.**
  Adenosine, Magnesium, Potassium

102

Drug Therapy of Cardiac Arrhythmias

Class I Antiarrhythmics

❖ This class of drugs block Na+ channels
❖ They reduce the maximum rate of depolarization (MRD) during phase 0
❖ They have negligible effect on resting cardiac membrane potential but inhibit AP in excitable cells → decrease conductivity & contractility
❖ They suppress both normal Perkinje fibres & His bundle automaticity resulting form myocardial damage
❖ Class I drugs bind to Na+ channel more strongly when they are in either open or refractory state

Class I B

❖ *Lidocaine, Phenytoin, Mexiletine & Tocainide*
MOA

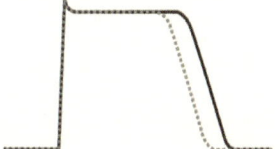

❖ Associates & dissociates fast from Na+ channels.
❖ The drug binds rapidly to open state of Na+ channels during phase 0 of AP, but dissociates rapidly too

❖ By the time the AP reaches its peak, it rapidly blocks other channels → development of premature beats will be suppressed because Na+ are already blocked rapidly.
❖ They bind selectively to refractory Na+ channels & blocks preferentially when the cardiac cells are depolarized. E.g., ischaemia & MI
❖ Nominal effect on the rate of rise of phase 0, shortens phase 3 repolarization → decrease ERP & duration of AP
❖ Shortening of phase 3 → decrease in QT interval or no effect
❖ Minimal effect on conductivity & contractility
❖ No effect on healthy myocardium & conducting tissue
Therapeutic Uses

❖ These drugs are used for Ventricular Tachycardia, PVCs & Ventricular Extrasystoles results from acute MI or Open Heart Surgery
❖ Phenytoin & Lidocaine are most commonly used to treat digitalis induced ventricular & Supraventricular arrhythmias
Lidocaine

❖ Most popularly & commonly used local anesthetic
❖ Given by IV infusion as a loading dose of 150-200mg for 15 min followed by a maintenance infusion @ 1-4 mg/min (not more than 300 mg in 1 hr)
❖ Used prophylactically for post MI case (basis not known) → morbidity rate is more → may be by increasing the incidence of asystole
❖ High 1st pass metabolism → dose adjustment needed in liver dysfunction patients & CHF
❖ Least cardiotoxic drug
❖ Extra-cardiac effects: drowsiness, convulsions, slurred speech, confusion & paraesthesia

Class I C

❖ *Flecainide, Encainide, Propafenone & Moricizine*
MOA

❖ Associates & dissociates slowly from Na+ channels →
steady state level of channel block during the cardiac cycle
❖ General reduction in conductivity & contractility
❖ Inhibit conduction through His-Purkinje system
❖ Decrease the rate of phase 0 depolarization in Purkinje &
Ventricular myocardial fibres → slowing of conduction in

all cardiac tissue with a minimal effect on ERP & action potential duration.

❖ These drugs reduce automaticity, decreases AV conduction & contractility
❖ Automaticity is decreased by a rise in threshold potential rather than a decrease in the
slope of phase 4 depolarization.
❖ They prolong PR & QRS intervals but not the QT interval
❖ They retard reentry of retrograde & antegrade impulse
Therapeutic Uses

❖ They suppress PVCs, Ventricular Tachyarrhytjmias, Supraventricular Arrhythmias.
❖ As they retard anterograde & retrograde conduction in the bypass tract of Wolf-Parkinson-
White Syndrome

Class I A

❖ *Quinidine, Procainamide & Disopyramide*
MOA

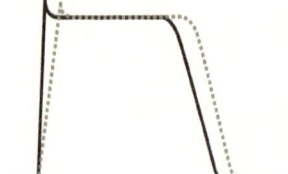

❖ Block open & inactivated Na+ channels but show an intermediate
association-dissociation properties which lie between I B & I C

❖ They slow the rate of phase 0 of the AP, increase the ERP, prolong
the action potential duration, moderately depress conduction,

increases QRS & QT interval

❖ Decrease the slope of phase 4 depolarization in pacemaker cells other than SA node
Therapeutic Uses

❖ Used to prevent acute, chronic ventricular & Supraventricular arrhythmias especially PSVTs,
PVCs, PACs, VT
Class II Antiarrhythmics

❖ *Propranolol, Acebutalol, Esmolol*
MOA

- ❖ β-Blockers → competitively block catecholamine-induced stimulation of cardiac β-receptors & depress phase 4 depolarization of pacemaker cells
- ❖ slow sinus & AV conduction → decrease in HR & prolongation in PR interval (atrial depolarization)
- ❖ they increase the ERP & prolong the duration of AP by decreasing conduction through AV node

Therapeutic Uses

- ❖ Propronolol, Atenolol → used to control Supraventricular Arrhythmias (e.g., Atrial Flutter or Fibrillation & PSVTs), treat Tachyarrhythmias caused by adrenergic stimulation (e.g., in hyperthyroidism & during Anaesthesia), suppress severe ventricular arrhythmias in prolonged QT syndrome, treat digitalis induced ventricular arrhythmias, terminates PVCs in patients without structural heart disease, treat atrial fibrillation & WPW syndrome
- ❖ Esmolol used to treat Supraventricular tachycardias, to control abnormal ventricular response to atrial fibrillation or flutter during or after surgery.

Class III Antiarrhythmics

- ❖ *Amiodarone, Bretylium, Ibutilide, Dofetilide, Sotalol*

MOA

- ❖ They block outward K+ channels during phase 3 of AP
- ❖ These drugs prolong the duration of AP without affecting phase 0 of depolarization or the resting membrane potential.

- ❖ They prolong the ERP and QT & PR intervals.

Therapeutic Uses

- ❖ Amiodarone → used to control life threatening ventricular arrhythmias
- ❖ For the treatment of ventricular fibrillation, ventricular tachycardia, Supraventricular arrhythmias
- ❖ Oral doses of 300-600 mg/day can be given as maintenance dose in ventricular tachycardia
- ❖ For IV infusion → a rapid loading dose of 150 mg is infused within 10min → followed by slow infusion of 1mg/min for 6hrs (total 360 mg) → then maintenance infusion of 0.5 mg/min for the remainder of 24 hour period.

Class IV Antiarrhythmics

- ❖ *Verapamil, Diltiazem*

MOA

- ❖ They block both activated and inactivated L-type Ca+2 channels in myocardium resulting in a decrease in the rate of phase 4 depolarization in SA node, AV node & slowed conduction through AV node.
- ❖ The drugs also prolongs the ERP of AV node
- ❖ A decrease in inward Ca+2 current also occurs during upstroke i.e. phase 0 of the AP
- ❖ Ca+2 channel blockade in SA node & AV node results in an increase in PR interval.
- ❖ Verapamil has a stronger action on heart than vascular smooth muscle.
- ❖ Nifedipine group of Ca+2 channel blockers exert stronger effects on vascular smooth muscles → are used as antihypertensive/antianginal agents
- ❖ Diltiazem exhibits mixed actions

Therapeutic Uses

❖ Both of these drugs mainly used to treat Supraventricular Arrythmias
❖ First line drugs for PSVTs stemming from AV nodal reentry
❖ They rapidly control the ventricular response to atrial flutter & fibrillation
❖ Verapamil → To control PSVTs → in a dose of 80-120 mg TID or QID
❖ To control Atrial arrythmias → 5-10 mg/hr continuous IV infusion till adequate response is obtained
❖ Diltiazem → to control PSVTs → 30-90 mg every 6hrs orally
❖ To control Atrial Arrhythmias → 5-15 mg/hr continuous IV infusion till adequate response is obtained.

Conduction Block

➢ Class I: retards conduction enough so that beat still gets through normal cardiac tissue but not through any weakened tissue
➢ Class III: prolongs refractoriness

Adenosine

✓ Not in Vaughan Williams class
✓ Purine nucleotide (activates adenosine receptors)
✓ Slows AV nodal conduction
✓ Acute Rx
✓ Termination of SVT/ diagnosis of VT
✓ Given IV only (rapid bolus)

Adenosine – Adverse Effects

✓ T1/2 < 2seconds
✓ Feeling of impending doom!
✓ Flushing, dyspnoea, chest pain, transient arrhythmias
✓ Contraindicated in asthma, heart block

Torsade De Pointes

✓ Long Q-T syndrome
✓ Polymorphic Ventricular Tachycardia
✓ IA drugs can also cause this
✓ Blocking of potassium channels and prolonging repolarization

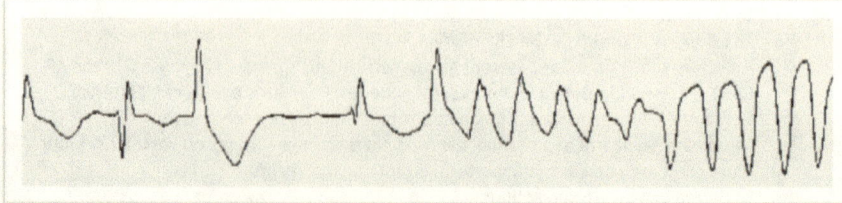

Choice of Antiarrhythmic Drugs for Some Commonly Encountered Arrhythmias

Atrial Flutter

- ✓ Class IA – Quinidine
- ✓ Class II – Propranolol, Esmolol
- ✓ Class III – Amiodarone*, Ibutilide*
- ✓ Class IV – Verapamil
- ✓ Misc – Digoxin*

Atrial Fibrillation

- ✓ Class IA – Quinidine
- ✓ Class II –Esmolol, Propranolol
- ✓ Class III – Amiodarone*, Dofetilide*
- ✓ Class IV – Verapamil
- ✓ Misc – Digoxin*, Anticoagulant Therapy
- ✓ Anticoagulant therapy → reduce the risk of stroke that is associated with atrial fibrillation

PSVTs

- ✓ Class II – Propranolol*
- ✓ Class III – Amiodarone*, Sotalol*
- ✓ Class IV – Verapamil*
- ✓ Misc – Adenosine

Ventricular Tachycardia

- ✓ Class I A – Quinidine*
- ✓ Class I B – Lidocaine
- ✓ Class III – Sotalol*, Amiodarone*
- ✓ Misc – Magnesium

Ventricular Fibrillation

- ✓ Class I B – Lidocaine
- ✓ Class III – Bretylium, Amiodarone*

Wolf-Parkinson-White Syndrome (WPW)

- ✓ Class I C – Propafenone*, Flecainide
- ✓ Class II – Propranolol*

✓ Class III – Amiodarone*, Sotalol

Torsade De Pointes

✓ Class II – Propranolol*
✓ Misc – Magnesium

Digitalis Induced

✓ Class I B – Lidocaine, Phenytoin
✓ Class II – Propranol
* indicates drug of choice for chronic therapy and other drugs serve as alternatives

Coagulants
Anticoagulants

Coagulants

- **Vitamin K**
    - K_1(from plants, Soluble fats): Phytonadione
    - K_2(produced by bacteria): Menaquinones
    - K_3(Synthetic): Menadione, Acetomenaphthone (Fat Soluble); Menadione sod bisulfate, Menadione sod diphosphate (Water soluble)
- **Misc:** Fibrinogen(human), Antihaemophilic factor, Ethamsylate

Vitamin K (Koagulation)

- Vitamin K doesn't involve in the direct clotting mechanism
- It is required for the synthesis of FOUR clotting factors from liver cells
- Factor II (Prothrombin)
- Factor VIII (Proconvertin)
- Factor IX (Christmas Factor)
- Factor X (Stuart Factor)

Mechanism of Action

- Vit K act as a cofactor at a last stage in the synthesis by liver coagulation proteins – prothrombin, factors VII, IX & X
- The Vitamin K dependent change confers on them à the capacity to bind Ca^{+2} & to get bound to phospholipid surfaces ß properties essential for participation in the coagulation cascade

Vitamin K Analogues

- **Natural Vitamin K or Phytonadione** à obtained from leafy vegetables
- It can be given orally or by IV
- It has rapid onset & prolonged effect
- When it is given orally à requires bile salts for absorption
- Vit K deficiency occurs in Obstructive Jaundice & in any disease involving malabsorption of fats
- **Vitamin K_2 or Menaquinone** à synthesized by intestinal bacterial flora
- Inhibition of the growth of intestinal bacteria from prolonged antibiotic therapy à Vit K deficiency
- **Vitamin K_3 or Menadione** à a synthetic anlogue
- Bile salts are needed for its absorption à it can be given orally in case of obstructive jaundice
- It has slow onset of action

Deficiency Vitamin K

- Manifested as increased bleeding tendency (e.g. postoperative bleeding) & hypoprothrombinaemia
- Greater risk of Vit K deficiency à New born & patients receiving anticoagulants à both have low levels of prothrombin & other clotting factors

Therapeutic Uses

- Prolonged Antibiotic use: any Vit K is given orally
- Overdose of oral anticoagulants: Vit K is specific dose, Vit K_1 is given i.m. à quick onset of action, Vit K_3 should be avoided à delayed onset of action
- Newborn or Premature infants: Vit K_1 1mg i.m. is administered
- Vit K_3 (Menadione should be avoided)

110

- Newborn or Premature infants: Vit K₃ (Menadione should be avoided) à it causes haemolysis & precipitates kernictures ß it competitvely inhibits glucuronidation of bilirubin
- Malabsorption syndrome/obstructive Jaundice: Vit K₃ orally or parenterally
- IV injection of Vit K à Dyspnoea, hypersensitivity

Systemic Procoagulants

Factor VIII products for treatment of Haemophilia A

- Antihaemophilic factor obtained from porcine or plasma
- Ultrapure anti-haemophilic factor à obtained from recombinant technology à it doesn't carry any risk of hepatitis or HIV transmission

Factor IX products for treatment of Haemophilia B

- Two preparations à Factor IX complex (contains significant amounts of factor II, VII, X) & purified factor IX recombinant

Mixed factor products for Haemophilia-A, vWD & Hypofibrinogenaemia

- Two preparations à Cryoprecipitated antihaemophilic factor (primarily contains factor VIII, fibrinogen & vWF [von Willebrand factor]) & anti-inhibitor coagulant complex (a mixed clotting factors preparation)

Desmopressin

- Synthetic desamino-arginine analogue of vasopressin

The advantages of desmopressin

- It retains the factor VIII & vWF incresing activity but lacks the vasoconstriction action of vasopressin
- No risk of HIV or Hepatitis
- It is administered parenterally or insufflated intranasally
- It is used to treat mild haemophilia-A & vWD

Ethamsylate

- It is a haemostatic drug
- It is believed to work by increasing capillary endothelial resistance and promoting platelet adhesion
- It also inhibits biosynthesis and action of those PGs which cause platelet disaggregation, vasodilation & increases capillary permeability
- **Uses** – in the prevention & treatment of capillary bleeding in menorrhagia, after abortion, epistaxis, malena, hematuria, after tooth extraction
- **S/Es** – Nausea, rash, headache, fall in BP
- **Dose:** 250-500 mg TDS Oral

Local Haemostatics (Styptics)

- **Thrombin:** obtained from bovine plasma
- **Fibrin Sealant:** obtained from human plasma
- **Gelatin Foams:** soaked with saline, fibrin or thrombin to cover bleeding area
- **Vasoconstrictors:** cotton gauze with 1% epinephrine + 0.5% oxymetazoline
- **Astringents:** Tannic acid (1%)
- **Russels viper venom**

Anti-Coagulants

Classification of Anticoagulants

- **Parenteral Anticoagulants**
  - *Heparin*
  - *Low Mol Wt Heparins:* Enoxaparin, Dalteparin, Tinzaparin, Ardeparin, Nadoparin, Raviparin
  - *Synthetic Heparin:* Fondaparinux
  - *Misc group:* Lepirudin, Bivalirudin, Argatroban, Danaparoid, Drotrecogin alfa
- **Oral Anticoagulants**
  - *Coumarin Derivatives:* Dicumarol, Ethyl biscoumacetate, Warfarin, Acenocumarol
  - *Indandione Derivatives:* Phenindione, Anisindione

Heparin

- It is a mixture of sulfated mucopolysaccharides with a molecular weight ranging from 10,000-40,000
- It is a strong electro-negatively charged acidic polymer à it has a number of anionic SO_4^- & COO^- acidic groups in its
- It is present together with histamine in all tissues containing mast cells
- Richest source – Lung, Liver, Intestinal Mucosa
- It is obtained from beef lung & pig intestinal mucosa

MOA of Heparin

- Antithrombin-III à blocks the activity of activated clotting factors XII, XI, X, IX & II ß this action is very slow ß action of Antithrombin-III is accelerated by Heparin against IIa, Xa
- Thrombin (IIa) & Factor Xa à sensitive to the inhibitory effects of High Molecular weight Heparin-Antithrombin-III complex à IIa inhibition is more than Xa
- Positively charged AT-III & Thrombin bind à Long chain heparin molecule à conformational changes in AT-III to expose à makes it more reactive site à accessible to IIa/Xa binding
- Heparin – AT-III – Clotting factor ternary complex forms à heparin dissociates and leaves the inactive AT-III – Clotting factors complex à free heparin goes and binds with newer AT-III
- Xa binds only with AT-III portion of long chain heparin-AT-III complex
- Low molecular weight – AT-III complex can inhibit only factor Xa, but not Factor IIa àReason: LMWH heparin is a very small molecule to bind at time with AT-III & IIa
- Heparin-AT-III complex is very efficient inhibitor of free thrombin but not the clot bound thrombin
- Heparin in higher doses à **inhibits platelet aggregation** [secondary action] à thrombin is a powerful platelet aggregating agent
- Heparin à also **has lipaemia clearing effect** à **makes the turbidity free** à it is due to facilitated release of lipoprotein lipase from vessel wall and tissues
- Its not absorbed from GIT, doesn't cross BBB, BPB
- Its given slow IV infusion or by deep SC not by IM à to avoid haematoma formation at injection site
- Low dose SC injection is preferred for prophylaxis 1-2 hrs before the surgery à bioavailability is inconsistent
- It is metabolised in liver by Heparinase à half-life increases with increasing dose
- Duration of action of heparin is short (1-5hrs)

LMW Heparins

- Unfractioned heparin (5000-40000) à on fractionation à heparin & LMWH
- LMWH have shorter polymer length à lesser effect on thrombin, platelet function, on coagulation
- Thrombocytopenia is less frequent, chances of haemorrhage are less

Advantages of LMWH*(Imp)

- These can be given SC, consistent & better bioavailability (70-90%)
- These have longer elimination half-life & eliminated by 1st order kinetics à effects are more consistent & dosing is less frequent (OD)
- These don't prolong APTT(Activated Partial Thromboplastin Time) & whole blood clotting time à response is predictable & monitoring is not required
- Dose is given in mg (not in units) & calculated on body weight basis

Differences from unfractioned Heparin*(Imp)

- **Average molecular weight**, heparin is about 20000 Da and LMWH is about 3000 Da
- **Once-daily dosing**, rather than a continuous infusion of unfractionated heparin
- **No need for monitoring** of the APTT **coagulation parameter**
- Possibly a **smaller risk of bleeding**
- **Smaller risk of osteoporosis** in long-term use
- **Smaller risk of heparin-induced thrombocytopenia**, a feared side-effect of heparin
- The anticoagulant effects of heparin are typically reversible with **protamine sulfate, while the effect on LMWH is limited**
- Has **less of an effect on thrombin** compared to heparin, **but maintains the same effect on Factor Xa**

Synthetic Heparin Derivatives

- **Fondaparinux:** it is a synhetic pentasaccharide
- It causes an AT-III mediated selective inhibition of Factor Xa à interrupts the blood coagulation cascade à leading to ultimately inhibition of thrombin formation
- It doesn't cause inhibition of thrombin directly due to its shorter polymer length
- Its given SC once daily (2.5mg), Bioavailability is 100%, time to peak – 2-3hrs, t1/2-17-21hrs
- Lesser antiplatelet, chances of thrombocytopenia are less
- It is preferred for thromboprophylaxis of patients undergoing hip or knee surgery & for therapy of pulmonary embolism

Heparin Antagonist (Protamine Sulfate)

- the use of heparin antagonist is rare because the duration of action of heparin being short ß discontinuation of heparin solves the problem
- In cardiac/vascular surgery à there needs to terminate heparin action rapidly
- Heparin overdose à protamine sulfate
- It is a basic low mol wt protein given IV à neutralizes strongly acidic heparin weight by weight
- The use of excess protamine should be avoided à itself has an anticoagulant effect
- Being a basic drug à it releases Histamine à hypersenitivity reaction, flushing, bronchoconstriction
- Neutralisation of LMWHby protamine is incomplete
- It can't antagonise the actions of Fondaparinux

Misc Parenteral Anticoagulants

- Most of them act by binding the thrombin directly à inhibit thrombin effects in the downstream of blood coagulation cascade

- They never bind to AT-III nor to other plasma proteins

Hirudin & Lepirudin

- Hirudin is a specific irreversible inhibitor of thrombin obtained from the salivary glands of leech
- Lepirudin is the recombinant form of hirudin
- It is used as anticoagulant in the patients with heparin-induced thrombocytopenia & associated thrombo-embolic disease
- Its unable to use is for prevention of ischemic complications associated with unstable angina
- It is given parenterally & monitered by APTT(Activated partial thromboplastin time)
- Adverse Effects – haemorrhage, haematuria & increased transaminase

Bivalirudine

- It has rapid onset of action but shorter duration
- It is used in percutaneous coronary angioplasty

Argatroban

- It is used in heparin-induced thrombocytopenia
- It has a short duration of action

Drotrecogin Alfa

- It's a human recombinant activated Protein-C
- It inhibits Va & VIIIa
- Its used to decrease the mortality from severe sepsis (associated with organ dysfunction) in adults at high risk of death

Danaparoid

- Its 84% heparin + 12% dermatan + 4% chondrotin à obtained from pig intestinal mucosa
- It inhibits activation of factor Xa which ultimately inhibits thrombin generation
- It is used for prevention of postoperative deep vein thrombosis following active hip replacement surgery & also for patients with heparin-induced thrombocytopenia
- $T_{1/2}$ – 24 hrs

Oral Anticoagulants

- They are relatively inexpensive, self adminisered oral anticoagulant therapy
- They are Dicumarols, Indanediones
- But Indanediones are more toxic than Dicumarols

Mechanism of Action

- They act by competitively inhibiting Vit K reductase à inhibit the synthesis of Clotting factors II, VII, IX, X by the liver
- The reduced form of Vit K (KH) à serves as cofactor in the γ-carboxylation of clotting factors
- In this carboxylation, Vit K (KH) is oxidised to its epoxide (KO) à which is reduced enzymatically by Vit-K reductase ß all these drugs inhibit this reduction
- Their anticoagulant effects take 1-3 days to develop à because Prothrombin is depressed at the last ß prothrombin half-life is 2-3 days à it will take time to decline the levels of prothrombin
- Monitoring of coumarin therapy is done by estimating prothrombin time of blood
- They don't alter bleeding time

Warfarin*

- Commercial Warfarin is a racemic mixture of R(+) & S(-) warfarin

114

- S(-) Warfarin is 4 times more potent then R(+)
- Oral bioavailability is 100%, Plasma protein bound 99% à Low V_d (7.7L), Longer half-life-36hrs & several drug displacement reactions
- It is metabolised in liver à undergoes enterohepatic circulation but partly conjugated with glucuronic acid
- Coumarins cross placental barrier, the active drug excreted through milk

Warfarin Drug Interactions

- Large number of drugs may interact with warfarin
- they increase the activity resulting à unexpected haemorrhage
- Decrease warfarin activity à therapeutic failure

Warfarin activity is increased

- **Enzyme inhibitors** such as Metronidazole, chloramphenicol, disulfiram, erythromycin, cimetidine, allopurinol, amiodarone, celecoxib
- **Drugs which displace warfarin from protein binding sites** such as Cotrimoxazole, indomethacin, phenytoin, probenecid, phenylbutazone, ethacrynic acid
- **Liquid paraffin:** due to emulsification & excretion of vit K à leading to vit K deficiency
- **Drugs which inhibit gut flora & reduce vi K synthesis:** broad spectrum antibiotics
- **Drugs causing hypoprothrombinaemia such as 3rd generation cephalosporins** like cefoperazone, cefamandole, moxalactam
- **Misc Factors:** Hepatic disease (decrease the synthesis of clotting factors), hyperthyroidism (enhances catabolism of clotting factors)

Warfarin activity is decreased

- **Enzyme inducers** such as barbiturates, rifampicin, griseofulvin, carbamazepine
- **Drugs that bind to warfarin & inhibit its absorption,** e.g., cholestyramine, sucralfate
- **Drugs that increases the synthesis of clotting factors,** e.g., Estrogen containing oral contraceptive pills
- **Misc Factors:** Hypothyroidism 9reduced catabolism of clotting factors), hereditary resistance

Warfarin Antagonist

- **Vit K (K1-Phytonadione)** à used to treat the overdosage toxicity of coumarin group of oral anticoagulants
- **Fresh frozen plasma or factor IX concentrate** à which contains large amounts of prothrombin levels à used in warfarin over dosage poisoning

Uses of Anticoagulants

- Anticoagulants don't dissolve the clot that has already been formed
- They prevent the thrombus extension, recurrence, embolic complications by reducing the rate of thrombin formation
- Heparin is used initially, for its rapid action à followed by a oral anticoagulant à Heparin is discontinued after 6-7 days when warfarin reaches its full therapeutic effects

Used to Prevention & in the Treatment of Deep Vein Thrombosis & Pulmonary Embolism

- Vein Thrombosis à fibrin thrombi [the formation or presence of a blood clot within a blood vessel during life] à anticoagulants are very effective
- Prophylaxis is usually needed for bedridden patients, old, postoperative cases, patients with hip/leg fracture à intermittent low dose heparin administration (5000 IU, SC, 8-12 hrly) or LMWH* (e.g., Enoxaparin 30mg SC 12hrly or Dalteparin 5000 U SC per day) [rationale]
- For established venous thrombosis à a bolus i.v. injection of 5000-10000 IU of Heparin followed by IV infusion of 100 U/hr

115

- Heparin is used for 6-7 days followed by warfarin with a 3 days overlap period
- Warfarin should be started with 10-15mg and reduced to 5-7 mg/day when plasma thrombin activity reduced to 25% of the normal value

Used in the Myocardial Infarction

- Arterial thrombi à due to platelet thrombi à antiplatelet drugs then anticoagulants
- But these can be used to reduce secondary thromboembolic complications
- LMWH à prescribed for shorter period till patients become ambulatory
- These may also be used during coronary angioplasty & stent placement

Used in the Unstable Angina

- angina pectoris characterized by sudden changes (as an increase in the severity or length of anginal attacks or a decrease in the exertion required to precipitate an attack) à to reduce the chances of MI in unstable angina, Aspirin + Heparin followed by Warfarin

Used in the Rheumatic Heart Disease

- If there are chances of thrombosis or embolism from atrial fibrillation à Warfarin or low dose heparin or low dose aspirin used to prevent the chances of stroke
- Warfarin is more effective in atrial fibrillation patients with high risk of stroke
- Aspirin is for low risk patients or those who can't take warfarin

Misc Uses of anticoagulants

- Treatment of disseminated intravascular coagulation and peripheral embolism (retinal vessel)
- During cardiac bypass surgery & in placing artificial heart valves to prevent formation of any thrombus/emboli
- To prevent clotting during haemodialysis
- For anticoagulation in pregnancy

Adverse Effects

- Haemorrhage is most common

A/Es of Heparin

- Thrombocytopenia
- Osteoporosis, Spontaneous fractures on prolonged use
- Transient Alopecia, Hypersenstivity reactions

A/Es of Warfarin

- Teratogenic effects à haemorrhage disorders in foetus & foetal bone deformities
- Transient Alopecia
- Dermatitis & Diarrhoea

Contraindications [Heparin]

- Bleeding disorders, thrombocytopenia, severe hypertension (risk of cerebral haemorrhage), threatended abortion, bleeding piles, subacute bacterial endocartitis (chances of embolism), tuberculosis (chances of haemoptysis), lumbar puncture

Contraindications [Warfarin]

- Bleeding disorders, thrombocytopenia, severe hypertension (risk of cerebral haemorrhage), threatended abortion, bleeding piles, subacute bacterial endocartitis (chances of embolism), tuberculosis (chances of haemoptysis), lumbar puncture
- Pregnancy à skeletal abnormalities in fetus
- Necrosis of soft tissues (buttocks & breast) [basis: due to thrombus formation in venules which results due to inhibition of biosynthesis of protein-C (a natural anticoagulant which activates factor V & VIII) à affected earlier resulting in precoagulant state à this is the rationale for the initial therapy of Heparin before Warfarin]

Q) Differences between Heparin & Warfarin (Dicumarol)*(Imp)

Heparin	Warfarin
✓ It is a parenteral anticoagulant	✓ It is a oral anticoagulant
✓ Anticoagulation action is by accelerating AT-III and inactivating Factors IIa & Xa	✓ Anticoagulant action is by inhibiting the Vit K reductase which indirectly inhibits the synthesis of coagulant factors
✓ Duration of action is shorter	✓ Onset of action is slower & Duration of action is longer
✓ Continuous monitoring of coagulant parameters in blood after administration	✓ Its not safer in pregnancy ☐ it causes skeletal abnormalities in fetus
✓ It is safer in pregnancy	✓ It also causes necrosis of soft tissues
✓ But it causes thrombocytopenia, haemorrhage	✓ Warfarin antagonist is Vit K₁-Phytonadione
✓ Heparin antagonist is Protamine Sulfate	✓ Heparin should be given before the warfarin administration to avoid necrosis of soft tissues
✓ They are given initially before the warfarin administration	✓
✓	✓
✓	✓
✓	✓
✓	✓
✓	✓
✓	✓
✓	✓
✓	✓
✓	✓
✓	✓